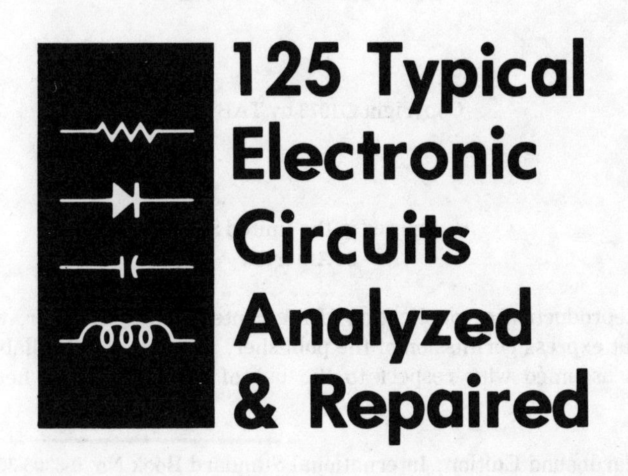

125 Typical Electronic Circuits Analyzed & Repaired

BY ART MARGOLIS

TAB BOOKS
Blue Ridge Summit, Pa. 17214

FIRST EDITION

FIRST PRINTING—JUNE 1973

Printed in the United States
of America

Hardbound Edition: International Standard Book No. 0-8306-3658-7

Paperbound Edition: International Standard Book No. 0-8306-2658-1

Library of Congress Card Number: 73-78197

PREFACE

When vacuum tubes reigned supreme in electronic gear, servicing thought processes centered around dc voltage readings. A B-plus voltage was produced by the power supply, and this voltage was thought of as the supply for the plates and screens of the tubes. A servicer took his trusty VTVM and searched along the path the voltage was dropped across. This was and still is a satisfactory first approach to troubleshooting a tube circuit.

Now that the solid-state device has become universal, this voltage tracing technique needs revised thinking. In a transistor an electrostatic charge on the control grid is not used to modulate cathode-to-plate electron movement. Instead, the emitter-to-base current flow varies the electron flow through a transistor. As a result, a supplemental mind's-eye view of electron movement is needed. The electron-movement thinking can be used in tube circuits, too. The important thing is to have a clear idea of the difference between the voltage-dropping and the electron-movement ways of looking at things.

Voltage is dropped from the B-plus power supply point along a path to the plate of a tube or the collector on an npn transistor. Electron flow goes the other way. Electrons leave the plate or collector and follow a path to their attraction point—the B-plus power supply point. In a pnp transistor, the power supply point is B-minus. Electrons leave B-minus and follow a path to the collector of the pnp. Once the idea of electron flow becomes second nature to the servicer, he has taken a giant step toward acquiring expert servicing capabilities.

The circuits in this book are analyzed with electron-flow thinking in mind, beginning with the simplest of circuits and gradually working up to the most complex, such as color TV.

The most logical path a servicer can take during a repair can be described, with a little practice, on a flow chart. At the end of the book are complete flow charts for one of the most complicated types of electronic gear—a solid-state color television receiver.

I'd like to thank Mr. R. C. Hannum, Supervisor, Training & Technical Publications of General Electric, for his help in providing the GE schematic and other service notes.

Art Margolis

CONTENTS

	THE SERVICER'S POINT OF VIEW	13
1	HEATING ELEMENT CIRCUIT	18
2	SERIES TUBE HEATER CIRCUIT	19
3	PARALLEL TUBE HEATER CIRCUIT	21
4	FILTER SHOCKER CIRCUIT	22
5	SIMPLE FILTER CIRCUIT	23
6	TIME-CONSTANT CIRCUIT	25
7	SIMPLE CHOKE CIRCUIT	26
8	SIMPLE AUTOTRANSFORMER CIRCUIT	29
9	SIMPLE TRANSFORMER CIRCUIT	30
10	TYPICAL RF TRANSFORMER	31
11	LR TIME-CONSTANT CIRCUIT (FLYBACK)	33
12	IMPEDANCE-MATCHING CIRCUIT	35
13	SIMPLE TANK CIRCUIT	36
14	PEAKING COIL CIRCUIT	37
15	BYPASS CAPACITOR CIRCUIT	38
16	LIGHT BULB CIRCUIT	39

17 VACUUM TUBE DIODE CIRCUIT 40
18 SOLID-STATE DIODE CIRCUIT 41
19 ZENER DIODE CIRCUIT 44
20 VARACTOR CIRCUIT 46
21 CAT-WHISKER DIODE CIRCUIT 47
22 TRIODE VACUUM TUBE CIRCUIT 48
23 TETRODE & PENTODE VACUUM TUBE CIRCUITS 50
24 TRANSISTOR CIRCUIT 52
25 GROUNDED-CATHODE CIRCUIT 58
26 GROUNDED-EMITTER CIRCUIT 61
27 GROUNDED-GRID AMPLIFIER 62
28 GROUNDED-BASE AMPLIFIER 63
29 CATHODE-FOLLOWER CIRCUIT 64
30 EMITTER-FOLLOWER CIRCUIT 66
31 RESISTANCE-COUPLED TUBE CIRCUIT 67
32 RESISTANCE-COUPLED TRANSISTOR CIRCUIT 74
33 IMPEDANCE-COUPLED CIRCUIT 77
34 TRANSFORMER-COUPLED CIRCUIT 79
35 FET COUPLING CIRCUITS 80
36 CLASS A TUBE AUDIO AMPLIFIER 81
37 CLASS A TRANSISTOR AUDIO AMPLIFIER 83

38 CLASS B PUSH-PULL AUDIO AMPLIFIER 84

39 CLASS B TRANSFORMER-COUPLED (PUSH-PULL) TRANSISTOR 86

40 CLASS B TAPPED SPEAKER (PUSH-PULL) 87

41 CLASS B CASCODE (PUSH-PULL) 87

42 PARALLEL AUDIO AMPLIFIERS 88

43 SIMPLE CODE PRACTICE OSCILLATOR 93

44 ELECTRONIC INDUCTANCE (REACTANCE TUBE) 94

45 ELECTRONIC CAPACITANCE 96

46 NEUTRALIZING CIRCUIT 97

47 SIMPLE VOLTAGE STEPUP 98

48 SIMPLE VOLTAGE DOUBLER 99

49 ALTERNATE VOLTAGE DOUBLER 100

50 ACTIVE FILTERING CIRCUITS 101

51 RECTIFIER PROTECTION CIRCUIT 103

52 RF AMPLIFIER RECEIVER CIRCUIT 104

53 DIODE MIXER CIRCUITS 106

54 DIRECT CONVERSION CIRCUIT 107

55 TRANSMITTER RF AMPLIFIER TUBE CIRCUIT 108

56 TRANSMITTER RF AMPLIFIER TRANSISTOR CIRCUIT 109

57 TIME-CONSTANT AM DETECTION 109

58 CRYSTAL OSCILLATORS 110

59 FREQUENCY MULTIPLIER 111

60 FREQUENCY ABSORPTION CIRCUIT 112

61 DIP METER CIRCUIT 113

62 CAPACITOR TESTER (DIP METER) 114

63 INDUCTANCE TESTER (DIP METER) 115

64 CW MONITOR 116

65 CARBON MICROPHONE CIRCUIT 117

66 DYNAMIC MICROPHONE CIRCUIT 118

67 TRANSDUCER (MICROPHONE) CIRCUIT 118

68 PLATE MODULATION 119

69 SCREEN MODULATION 120

70 FM AND PM MODULATION CIRCUIT 121

71 DISCRIMINATOR CIRCUIT 123

72 PREEMPHASIS & DEEMPHASIS CIRCUIT 125

73 LIMITER AMPLIFIER 126

74 RATIO DETECTOR AMPLIFIER 127

75 METAL DETECTOR CIRCUIT 128

76 TRIGGER CIRCUIT 130

77 PROXIMITY CIRCUIT 131
78 PROXIMITY BRIDGE CIRCUIT 132
79 FET PROXIMITY CIRCUIT 133
80 SIREN CIRCUIT 134
81 FM BUG 135
82 DEBUGGING CIRCUIT 136
83 ILLEGAL JAMMER CIRCUIT 136
84 TELEPHONE TAP CIRCUIT 137
85 TELEPHONE FM BUG 138
86 TUBE TUNING INDICATOR 139
87 PHOTO DETECTOR CIRCUIT 140
88 SCR SWITCH CIRCUIT 141
89 SCR LOAD ADJUSTER 142
90 LIGHT DIMMER CIRCUIT 143
91 FLASHER CIRCUIT 145
92 MODEL CAR SPEED CIRCUIT 146
93 CRT GUN STRUCTURE 146
94 SHADOW MASK STRUCTURE 147
95 DEFLECTION YOKE CIRCUIT 148
96 HORIZONTAL CONVERGENCE CIRCUIT 150
97 VERTICAL CONVERGENCE CIRCUIT 151

98 and 99 VERTICAL SWEEP 152

100 HORIZONTAL OSCILLATOR 155

101 HORIZONTAL DRIVER 156

102 HORIZONTAL OUTPUT 157

103 VERTICAL BLANKING 158

104 HORIZONTAL BLANKING 160

105 DC RESTORER 160

COLOR TV RECEPTION 161

106 COLOR I-F CIRCUIT 162

107 COLOR KILLER CIRCUIT 163

108 BURST AMPLIFIER CIRCUIT 165

109 COLOR PHASE DETECTOR CIRCUIT 165

110 COLOR OSCILLATOR CIRCUIT 166

111 COLOR DEMODULATOR CIRCUITS 167

112 COLOR-DIFFERENCE CIRCUITS 168

113 TV FIRST I-F CIRCUIT 169

114 SECOND I-F CIRCUIT 170

115 THIRD I-F CIRCUIT 170

	AGC-SYNC CONFIGURATION	171
116	AGC KEYER CIRCUIT	172
117	AGC AMPLIFIER CIRCUIT	173
118	AGC DELAY CIRCUIT	174
119	SYNC SEPARATOR CIRCUIT	175
120	NOISE GATE CIRCUIT	175
121	NOISE GATE DRIVER CIRCUIT	176
122	TRANSFORMERLESS AUDIO OUTPUT	177
123	HALF-WAVE ANTENNA CIRCUIT	178
124	AUTOMATIC DEGAUSSING CIRCUIT	179
125	AUTOMATIC BRIGHTNESS CIRCUIT	181
	FLOW CHARTS	182
	APPENDIX I	184
	INDEX	203

THE SERVICER'S POINT OF VIEW

When the electronic servicer arrives on the scene, he is a "witness after the fact." The trouble has occurred. His job is to analyze the symptoms of trouble, form a theory on what has happened, and proceed along a logical path till he roots out the trouble. This mental exercise is the hard part of the repair. Once the defective part or connection is discovered, its repair or replacement is straightforward. Any mechanically inclined person with a little practice can change most electronic parts.

SERVICING EXPERTISE

Troubleshooting skill is needed in the "electronic detective" part of the job—the solving of the mystery, "What is causing the trouble?" When the experienced servicer looks at a circuit, he sees not a jumble of resistors, capacitors, coils, tubes, etc., but a familiar configuration that can do a job. He has a point of view. Each individual talented servicer has his own point of view. While some differences exist, they all originated from an understanding of the basics of circuits.

The servicer needs to use practically no mathematics. Expert servicing ability develops as one is able to visualize the movement of electrons in the circuit. When you correctly picture in your mind's eye what the electrons are doing in the circuit, you have made the giant step toward expert servicing.

BASIC CIRCUIT PROPERTIES

There are only three basic properties in an electronic circuit. They are resistance, capacitance and inductance. The properties are found in the resistors, capacitors, coils, vacuum tubes, pieces of n and p materials, and so forth. The components impede electron movement, store electrons, cause the electron movement to produce a surrounding magnetic field, and pass the electrons through a vacuum or adjoining pieces of semiconductors.

When you turn on a circuit the electrons start moving. They move because when you turn on the circuit you cause a

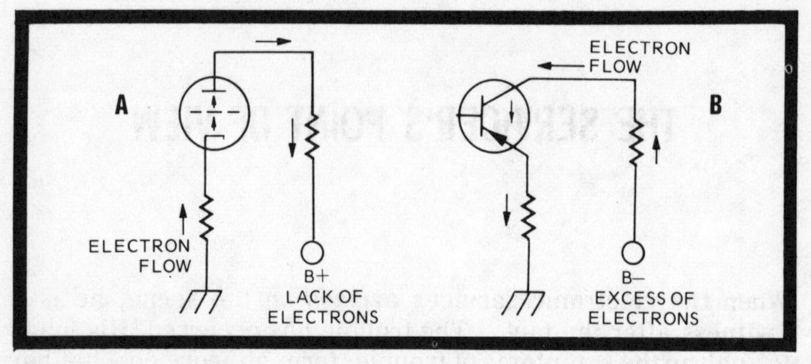

Fig. 1. Electron flow in all devices goes from an excess (–) to a deficiency (+).

battery or some other power source to produce a positive or negative charge somewhere in the circuit. A positive charge is a **lack of electrons**. A negative charge is an **excess of electrons**. Since electrons themselves are nothing but tiny negative charges, they are attracted to the positive charge and repelled by the negative charge.

When you turn on the circuit and insert the positive end of the battery in the circuit, the electrons come climbing up out of the ground and rush to the positive point (Fig. 1A) where the deficiency of electrons is attracting them. Should you turn on the circuit by placing the negative end of the battery in the circuit (Fig. 1B), the electrons pour out of the battery and repel the electrons that are lying dormant in the circuit.

The servicer, as he looks at the circuit, should immediately notice whether the B voltage is plus or minus. He almost sees the electrons running from ground to B-plus; or, if it is B-minus, the electrons are passing from the battery to ground. The direction the little negative charges move is designated on the schematic in that way.

Testing Electron Flow

The most important servicing technique is the test for proper electron flow. The circuit is examined and the nature of the flow is analyzed. If the flow is as it should be, the circuit is good. If the flow is incorrect, stopped, or flowing too heavily, a defect is indicated.

The most common way to check electron flow is with a voltmeter. The voltage at a particular test point depends on the deficiency or excess of electrons at that point. If a voltage is positive, it is attracting electrons; if negative, it is repelling electrons.

In a circuit, a positive voltage is normal when it is attracting the correct number of electrons. Should it start getting more electrons than it is designed to receive, the excess negative charges cause the voltage to lose its positive charge. As a result, the voltage drops. On the other hand, if the test point starts getting fewer electrons than it should, the lack of negative charges increases the positive charge and the voltage rises.

What could cause these changes? If a resistor in the circuit decreases in value (Fig. 2), it lets more electrons through and drops the voltage at the test point. Should that same resistor increase in value, fewer electrons will get through and the positive charge increases. Consequently, the voltage rises. While the voltmeter does not tell you how the electron flow is doing directly, by measuring the voltage at a test point and then drawing conclusions from the lowering or rising of the voltage, the resistor becomes a prime suspect. It can be the hidden defect in the circuit. This is a very simple example, but it is vital that you consider the electron flow in that manner as you gaze at a circuit.

Yes, you can test electron flow with an ammeter and get direct current readings. However, most TV schematics show the voltages at test points. Also, a current reading is time-consuming since the ammeter must be installed in series with

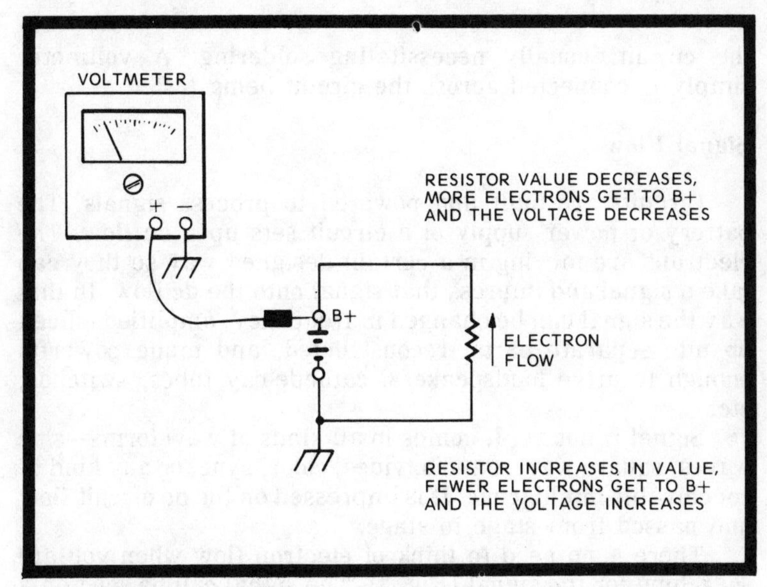

Fig. 2. Resistance stops electrons from getting to their destination, increasing the deficiency (+).

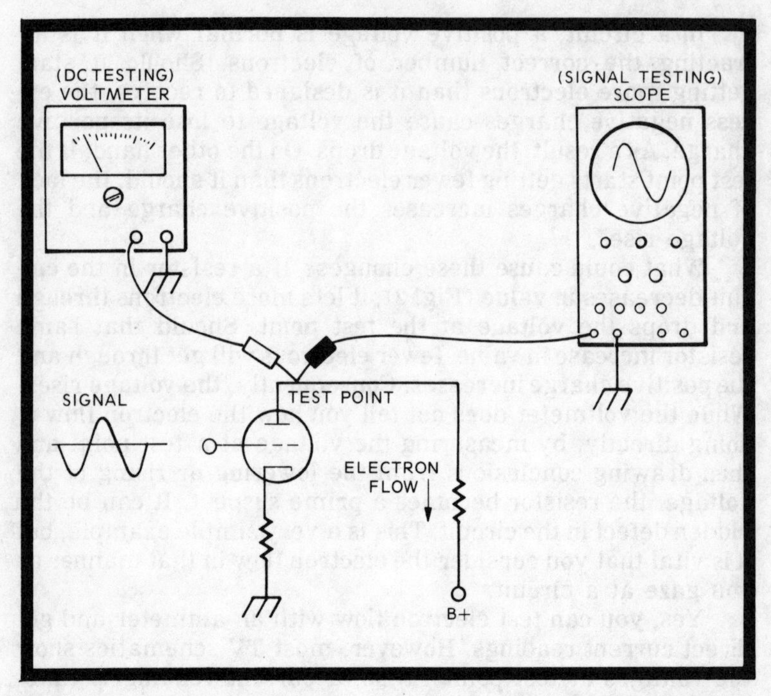

Fig. 3. A voltmeter measures dc potential. A scope reveals the peak-to-peak amplitude.

the circuit, usually necessitating soldering. A voltmeter simply is connected across the circuit being tested.

Signal Flow

Circuits are built and powered to process signals. The battery or power supply of a circuit sets up a dc flow. The electrons are moving in a certain designed way so they can take a signal and impress that signal onto the dc flow. In that way the signal can be changed in frequency, amplified, sliced up into separate parts, reconstructed, and made powerful enough to drive loudspeakers, cathode-ray tubes, switches, etc.

Signal is not dc. It comes in all kinds of waveforms—sine waves, square waves, audio, video, color, sync, or any kind of conceivable intelligence. It is impressed on the dc circuit flow and passed from stage to stage.

There is no need to think of electron flow when you are searching for the signal (Fig. 3). The signal can be seen on a scope, its amplitude read on the peak-to-peak (p-p) scale of a scope or p-p voltmeter, or another signal can be substituted in its place. Those are the tests to determine the presence of (or the amount of) signal distortion.

Much confusion reigns between the examination of test points for dc flow and signal flow. The servicer must have a clear distinction in his mind of the differences. Therefore, it must be realized that there are two separate functions in an electronic circuit: One, the circuit has its powering function. Some outside power source sets up a positive or negative potential. Electrons flow to the positive potential and away from the negative. The electrons are pressured to move as long as the circuit is closed and the potential exists.

The second function of a circuit is to handle signal—processing it or even producing the signal. The signal can be simple, such as a sine wave that comes out of an oscillator, or a composite color TV signal that is produced with the aid of cameras, microphones, and sync generators. The signal can pass through properly designed circuits, much like a train can pass through a railroad yard. Just as the train can be sidetracked, rerouted, slowed, shortened, and lengthened, so a signal can be similarly affected. Just as you wouldn't confuse a train with its tracks, don't confuse the signal with the electron flow.

1 HEATING ELEMENT CIRCUIT

The simplest traditional circuit is a power source and a resistor (Fig. 4). In a laboratory experiment you can take a flashlight battery and a resistor. The resistor is attached across the battery. As the circuit is closed current flows; the amount can be computed quickly with Ohm's old familiar law: voltage divided by resistance equals current.

The simplest circuit is also one of the most common. It is found in practically every electrical appliance from a toaster to a clothes dryer. Its value lies in the fact that when you pass current through a resistance, the temperature of the resistance rises.

In appliances, the unit is plugged into the outlet and the switch is closed. The 60 Hz current enters the circuit and electrons flow through the heating element. The resistance of the element causes the element to become red hot. (The element works exactly the same way whether ac or dc is passing through it.)

Resistance Circuit Failure

What could conceivably happen to such a simple circuit? Assuming current is entering properly, the only possible problem could be a change in resistance. The change could be anything from a dead short , which brings the resistance down near zero, to an open circuit, which causes infinite resistance.

If the circuit opens, the appliance goes dead; no electron flow occurs. The repair is easy. An ohmmeter is used to test the elements, the switch, the connections, and the line cord. When the open is found, it is repaired, either by replacement or by resoldering.

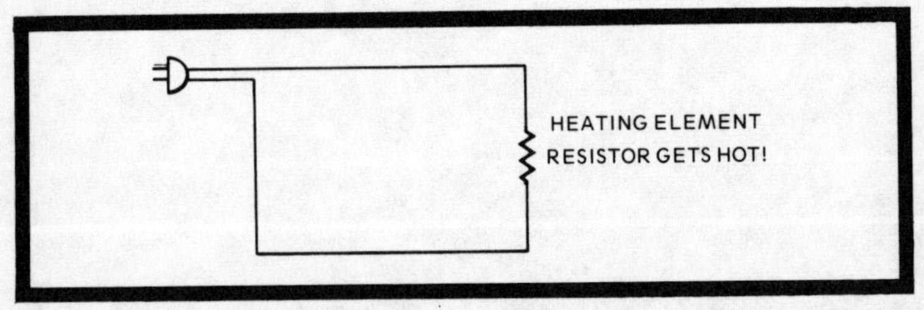

Fig. 4. The simplest circuit is used in lots of appliances. It's a resistor or heating element across the supply.

When the circuit develops a short, its total resistance drops and the amount of current flow rises. If the short is down near zero ohms, a large amount of current is drawn. If the current exceeds the power line fuse rating, the fuse will blow. A short in an appliance can easily draw enough current to exceed a fuse rating. Suppose the house fuse is rated at 15A. Since I equals E divided by R and E is 120V, if R becomes 3 ohms, the current (I) out of the fuse box becomes 40A—almost three times the load capability of the fuse!

A short circuit can be uncovered with the aid of the ohmmeter, too. The appliance cannot be plugged in with a short, so voltage readings cannot be made. The technique to locate the short is simple—valuable in all kinds of circuits—yet is ignored by many servicers. It's the technique of **disconnection**. It's ignored because it's troublesome. The connections must be unsoldered. It's not too bad in an appliance, but on a printed board it's tricky. However, it is an extremely useful technique.

The connections at the line cord, switch, and resistive element are disconnected one at a time. Resistance readings are made from the various points that were disconnected until the short is found. It is usually a shorted line cord or heating element. When the short is found, the defective part is repaired or replaced. Keep Ohm's law handy; it is used continually by the servicer.

SERIES TUBE HEATER CIRCUIT 2

The next step up after the simple appliance-type circuit is found in old table model radios that use tubes. The heaters in the tubes are resistors with a very low value, all hooked together in series. A typical heater string has a 35W4, 50C5, 12BA6, 12BE6 and 12AV6 (Fig. 5). These heaters can be powered by ac or dc; however, the voltage must be around 120 because each resistance has a **voltage drop**.

In order for the tube heater to do its job, enough electrons must push through it to make it red hot. That way it heats the cathode and causes the tube to work. The string is carefully designed so 120V will push enough electrons through the string to do the job.

The numbered prefix in such tubes typically tells the amount of voltage the tube needs to get hot. There is 35 + 50 + 12 + 12 + 12 = 121V. 120V is present at the 35W4. The voltage at the 50C5 drops to a value of about 80V. The 50C5 filament

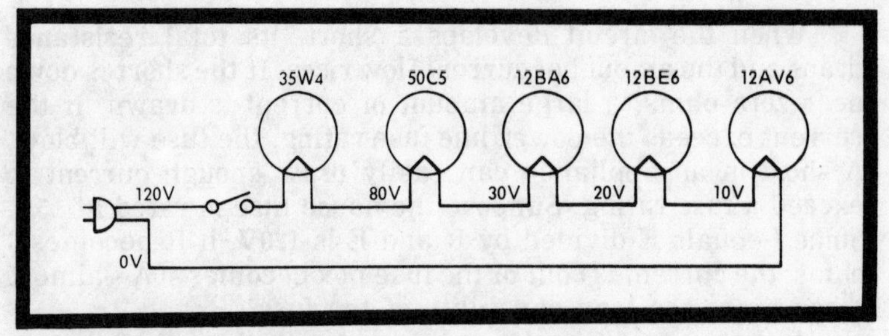

Fig. 5. In a series tube heater string each filament introduces a certain voltage drop, depending on its resistance.

drops the voltage to 30 at the 12BA6. Then it passes through the three 12V tubes and the circuit is returned to source at 0V or ground level.

Series Filament Circuit Failure

As in the appliance circuit, the only thing that can happen in the heater circuit is a change of resistance. Then the circuit does not operate as designed.

The most common trouble is an open, in which case the resistance becomes infinite. If you consider Ohm's law, when R is infinite I becomes zero, so no current flows. Usually, one of the tube heaters opens. Testing the tubes one by one reveals the bad one. If it's not a tube, it could be the power switch, the line cord, or a connection. In some series circuits, the tube filament voltage ratings do not add up to 120, so a resistor is installed to drop the voltage difference. The resistor is also a prime suspect in an open heater string.

Another trouble can be a short to ground in the middle of the string. Then the entire voltage drop is distributed through the heaters of the tubes that happen to be before the short (Fig. 6). For instance, suppose the third tube (12BA6) develops a heater-to-cathode short. All the voltage will be dropped in the 35W4 and 50C5. They will be overly bright and the other tubes dead. The two lit tubes will be carrying a lot more current than they were designed for. If you consider Ohm's law, R becomes smaller than the design value and I becomes higher. The tube heaters use up a lot more electricity when shorted than when operating properly.

With the resistances in series, an increase in resistance lowers the amount of current drawn and a decrease in resistance increases the amount of current drawn. The current present in the circuit is the same throughout, but the voltage drops from resistance to resistance.

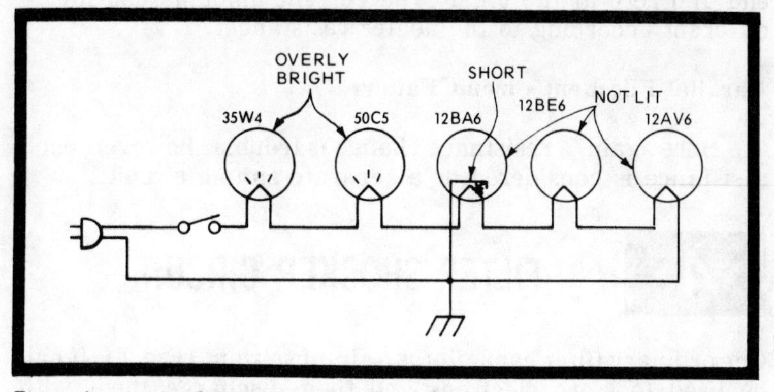

Fig. 6. In a series filament string, if a defect diverts electron flow, some of the tube heaters are bypassed.

PARALLEL TUBE HEATER CIRCUIT 3

A parallel tube heater string illustrates the type of circuit with resistances in shunt (Fig. 7). Most old tube-type car radios used this circuit. Typically, the car has a 12V battery. Therefore, radios were made containing all 12V tubes. That way the tube heaters could all be connected directly across the battery.

Resistances in parallel do not work together. In fact, they are almost completely independent of each other. Each is placed across the battery as a separate simple circuit. In order to figure out how much current a heater string draws, the current consumed by each heater is computed and then they are all added together. If the battery is good and the string does not draw more current than the battery is rated to deliver, the voltage is the same at each resistance – 12V at one

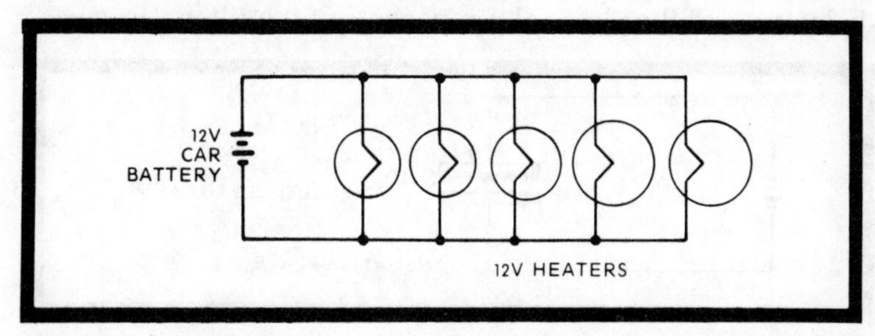

Fig. 7. In a parallel heater string each resistance requires the same voltage.

end and zero at the other. The current amounts can all be different according to the heater resistance.

Parallel Filament Circuit Failure

Here again, a resistance change is trouble; however, each resistance is considered as a separate simple circuit.

4 FILTER SHOCKER CIRCUIT

One ordinary filter capacitor is a simple circuit (Fig. 8). It can be used to store electrons and then discharge them with considerable shock-voltage force. This is handy to clear flaking inside the gun of a cathode-ray tube (Fig. 9).

A filter is usually a polarized capacitor in the microfarad range, termed an **electrolytic**. Since capacitance is a function of the thickness of the dielectric, having a dielectric only a few molecules thick produces great amounts of capacitance. (This can be compared with the relative thickness of paper, mica, or Mylar capacitors, normally in the picofarad range.)

The circuit consists of a battery and a filter. The battery is attached with correct polarity to the filter. As soon as the circuit is closed, electrons from the negative battery terminal flow to the negative plates of the filter.

The dielectric stops the dc flow of electrons and they pile up on the negative plate, constituting a negative voltage charge. This charge further repels the electrons on the positive plates. The lack of electrons constitutes a positive voltage charge. This activity keeps going till the capacitor attains the same voltage as the battery. When the battery is disconnected, the filter remains charged at the battery voltage because an electric field has been built up between the plates. Actually, the charge will soon leak off through the dielectric and the physical components of the filter. However,

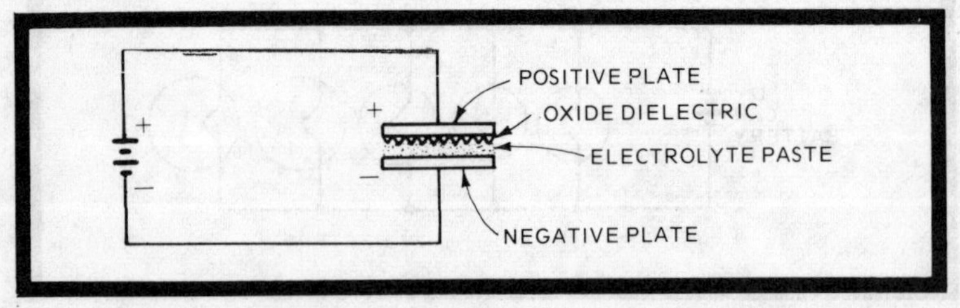

Fig. 8. A filter capacitor can be charged with a large voltage because it stores electrons on its negative plate.

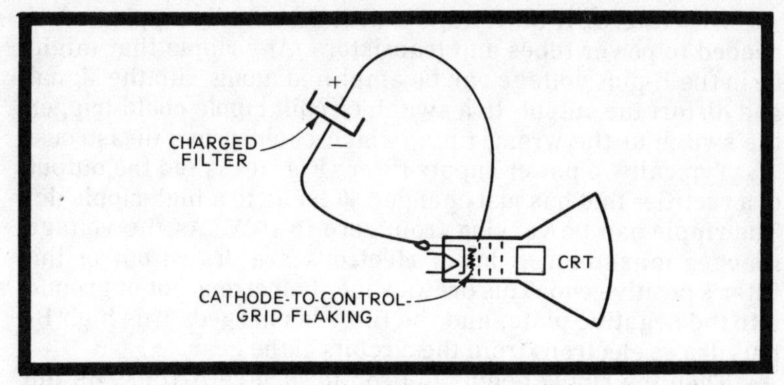

Fig. 9. A charged filter can be used to clear flaking between tube elements in a vacuum.

some capacitors have been known to keep the charge in storage for long periods of time.

During the time the filter was charged, current was flowing. Electrons also will move when the filter is finally discharged. This brings out the fact that current flows through a capacitor during charge and discharge, but there is no current flow through the capacitor when it is charged. All this statement means is that ac can pass through a capacitor, but dc is blocked off.

The filter shocker, once it is charged up, can be placed across the crt pins that lead to two elements in the gun which are clogged up with flaking. A circuit is formed through the gun, and the electrons leave the negative plates, pass through the gun, bridge across the flaking and go to the positive plates. A spark is seen inside the crt and the flaking is burned away.

Capacitor Circuit Failure

The filter can fail if holes develop in the dielectric. Then it shorts. Or the plate connections can break. Then it is open.

SIMPLE FILTER CIRCUIT 5

If a number of filters are connected in parallel, their capacitance adds together directly and a large total capacitance is obtained. However, the working voltage of this hookup is like a chain—no stronger than its weakest link. The working voltage of the group is no better than the working voltage of the lowest rated filter.

Why filters? A dc voltage that is essentially ripple-free is needed to power tubes and transistors. Any ripple that might be in the B-plus voltage can be amplified along with the signal and distort the output. In a switch circuit, ripple could trigger the switch at the wrong time, which could prove disastrous.

Typically, a power supply filter (Fig. 10) is fed the output of a rectifier that has just changed 60 Hz ac to a high-ripple dc. The ripple can be varying from zero to 100V. As the voltage reaches maximum, a lot of electrons are drawn out of the filter's positive end. This draws a lot of electrons out of ground into the negative plate, and the filter is charged. This high B-plus draws electrons from the circuits of the gear.

Then the ripple begins to drop toward zero volts. As the voltage recedes, the filter starts to discharge. However, a filter has a tremendous storage capacity due to its large capacitance. The amount of discharge is only a tiny percentage of the total charge. Meanwhile, the ripple is back up to 100V before the filter discharges to a percent or two of the total charge. So even though the ripple is sailing smoothly between 100V and zero, the capacitor is charged on the first few ripples and stays charged. The discharge time is so long that you can consider the filter as maintaining a charge up near the maximum ripple voltage. The filter acts almost like a battery. Once it gets charged, it maintains the voltage at near maximum.

Filter Circuit Failure

Filter capacitors can short, leak, or open. When one shorts in a B-plus line, the capacitor becomes a resistance. If the resistance is near zero ohms, a large current flow results. In

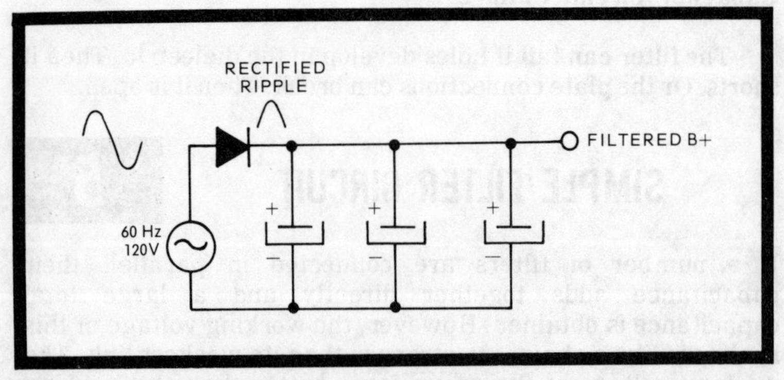

Fig. 10. A simple parallel filter stores a charge and is able to smooth out rectified ripple to dc.

properly fused circuits the fuse will blow. The short is quickly located with an ohmmeter, then the suspect filter must be disconnected and tested.

When a filter leaks, it acts as a filter with a resistor in shunt with it. The leakage resistance can be high or low. The higher it is, the less the trouble will be. All filters have some leakage. The lower the resistance, the worse the trouble will be and the filtering ability will decrease in direct proportion.

The most common trouble is an open filter, and it's the easiest to test. You simply bridge a known good filter across the suspect, carefully observing polarity. When you bridge the defective one and turn on the equipment, the gear will be restored to normal working order.

TIME-CONSTANT CIRCUIT 6

In electronic gear, when a resistor and a capacitor are placed together they form a charge-discharge network. If you multiply the resistance in ohms times the capacitance in farads, the result is the time constant in seconds.

This time is all important in setting up bias for tubes and transistors and all kinds of other circuits. The important thing to know is that the time constant is directly proportional to the resistance and capacitance. If you want to increase the time, increase either (or both) of these two values. Should you need to decrease the time, decrease the values.

Here is how an RC time-constant circuit works, whether it is shaping a vertical sync pulse or making an electronic organ play a note. There are four components in the RC timing circuit: a resistor, capacitor, switch, and voltage source (Fig. 11). When the switch is closed, a maximum amount of electrons leave the negative end of the battery and head in the direction of the positive end. The capacitor starts storing electrons and the resistor limits the flow. The resistance tends to increase the time the electrons take to flow to the capacitor plates. The size of the capacitor is important. The bigger the storage capability the longer it takes the capacitor to charge.

As the capacitor starts to store electrons, those electrons constitute a charge and tend to repel additional electrons that are trying to get into the capacitor for storage. To be exact, the time constant is the amount of time it takes for 63 percent of the battery voltage to appear on the capacitor. Once the RC network is charged, the time constant for charging is determined.

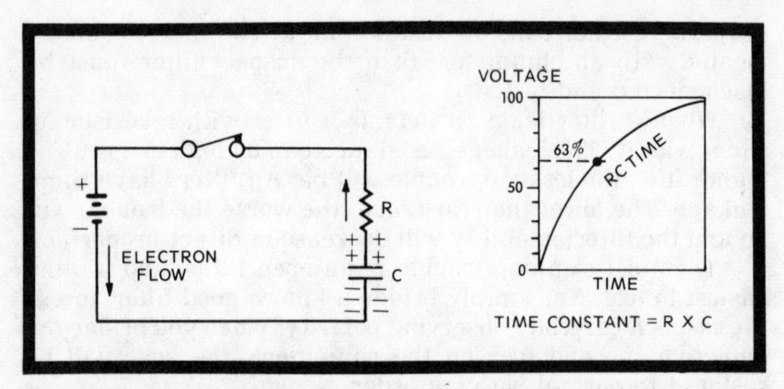

Fig. 11. The time constant of a resistor-capacitor circuit is the interval required for the capacitor to charge up to 63 percent of the input voltage.

When it comes time for the discharge to take place, the time constant comes into play again. To discharge the RC circuit, the battery must be removed. The switch is kept closed and the electrons stored on the negative plate of the capacitor flow through the switch and resistor to the positive plates as the voltage drops. The discharge time constant is the time it takes for the capacitor to drop its voltage to 37 percent of its charge.

RC Circuit Failure

Loss of the design time constant occurs when either the resistance or capacitance values change. The resistor can either increase or decrease in value. If it increases, the time constant increases. If the resistance decreases, the time constant decreases. If the resistor opens, the time becomes infinite. The capacitor can short, open, or leak. When this happens the time constant is destroyed, since the capacitor becomes a resistor.

7 SIMPLE CHOKE CIRCUIT

As electrons travel through a piece of wire, their movement creates an electromagnetic field around the wire (Fig. 12). This is easily proved with a compass. If a sine wave is passing through the wire, the number of electrons flowing varies in proportion to the amplitude of the voltage (Fig. 13). At 0V no electrons are forced through. At the 90-degree point (maximum voltage), a maximum number of electrons is being pushed into the wire. From zero to 90 degrees, the field builds. From 90 to 180 degrees, the field shrinks. The number of electrons passing through determines the size of the magnetic field.

As the magnetic field builds, a second voltage force is induced. This induced voltage is called the back voltage (Fig. 14). It opposes the forward voltage in the circuit and tries to stop the current from rising rapidly. An actual voltage drop takes place in the wire.

This opposition to the forward voltage is called **inductance**, since it is induced by the charging forward voltage. Inductance, measured in henrys, tends to stop the rise of the forward voltage by inducing the back voltage. Also, as the sine wave falls from an amplitude of 90 degrees back to 180 degrees, the energy that is stored in the back voltage is returned to the circuit and tends to slow down the voltage amplitude change. Since the inductance opposes the rise and fall, it is trying to smooth out or filter the variations of the sine wave. It is used in power supply filter circuits as a choke. First, the choke stops the rise by storing energy and then the choke stops the fall by giving the energy back. The electrons flow in a smooth, steady stream rather than in varying bursts.

It is found that if the wire is coiled, the magnetic field reinforces itself and produces more inductance. Other things that change the field are the radius of the coil and its length. If the coil is wound on an iron core, the magnetic field flux is trapped in the coil and it exerts a much stronger influence on the electron flow. This increases the inductance by factors in the hundreds. A typical iron core with good permeability increases the inductance 500 times.

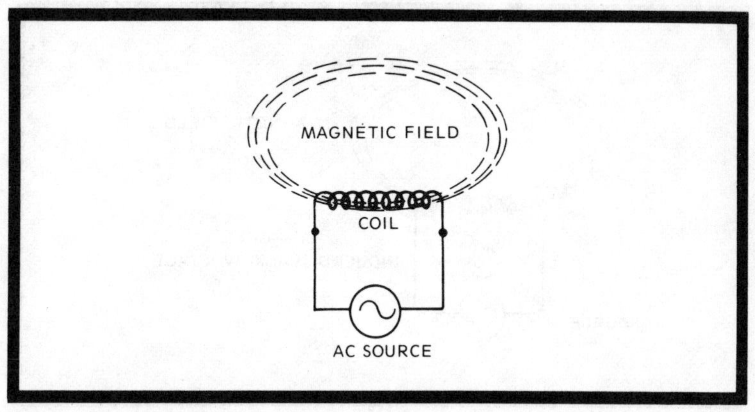

Fig. 12. As electrons flow through a coil, they build up a magnetic field around the coil.

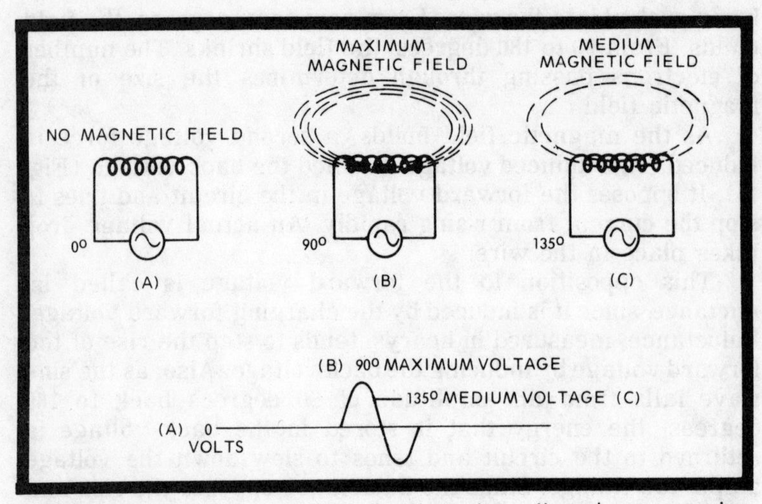

Fig. 13. A sine wave of current passing through a coil produces a varying magnetic field.

Choke Failure

It is rare to find a defective choke. On occasion you may find one with shorted turns, which sometimes occurs as the insulation between turns deteriorates with age (Fig. 15). Also, one of the turns can short to the frame of the choke. In 120V circuits the trouble is visually nailed down as the choke smolders. More often, a choke opens due to aging of the connections or when too much current is pushed through it. Power supply chokes are a luxury usually found only in circuits where expense is not too much of a problem.

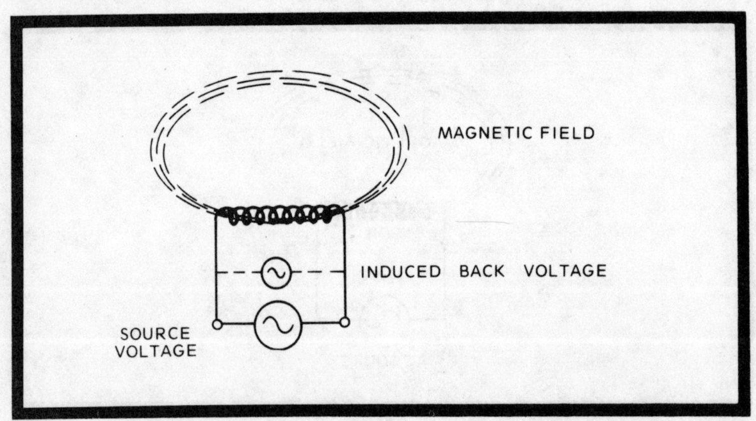

Fig. 14. As the magnetic field changes, it induces electron flow in the opposite direction of the original flow. It's called a "back voltage."

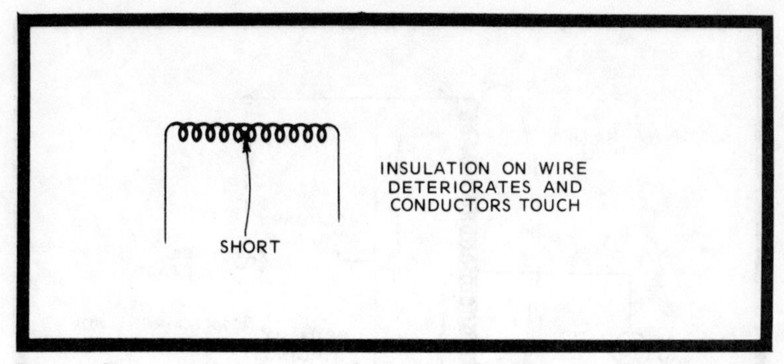

Fig. 15. The most common coil failure is a breakdown in the insulation between turns.

SIMPLE AUTOTRANSFORMER CIRCUIT 8

As noted in the choke circuit, as a sine wave passes through a coil, a magnetic field develops in direct proportion to the incoming changing voltage. Then the field produces a second back voltage that opposes the original voltage.

If you inject the voltage at a center tap and the bottom end of the coil (Fig. 16), the current flow occurs between these two connections. The magnetic field is induced around the coil. According to the strength of the field, it then induces the back voltage. However, since the field is in the air around the entire coil, it induces the back voltage in the entire coil, not just the part between the two injection points (the centertap and bottom).

If you then connect to the top and bottom of the coil, you get an output from the entire coil. The voltage output (assuming the amounts of current are small) will be about twice the value of the voltage input. If you had put 100V in, you'll get about 200V out. This is an **autotransformer**.

Autotransformer Failure

Autotransformer defects are shorted turns or an open in the winding. They are tested with an ohmmeter for resistance. An open coil is easy to find, but shorted turns are another matter. The only way the shorted turns can be detected is with a grid-dip meter. It applies a signal to resonate the coil. (The grid-dip meter coil is held close to the coil under test.) If the coil won't "ring," it has shorted turns.

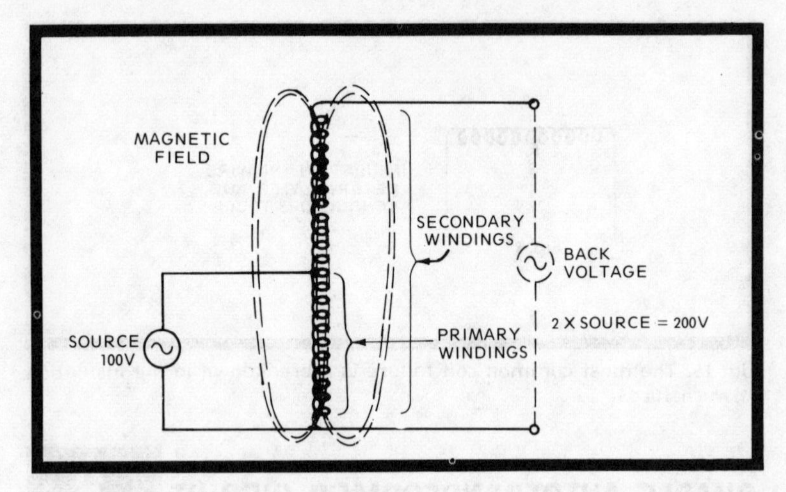

Fig. 16. In an autotransformer the primary and secondary windings are continuous with a tap located at the required point for the desired stepup or stepdown ratio.

SIMPLE TRANSFORMER CIRCUIT 9

In a conventional transformer, the primary and secondary are individual windings with complete dc isolation between the two (Fig. 17). Thus, different dc potentials can exist in the primary and secondary. An ac voltage in one winding will induce a proportional voltage in the other winding due to the mutual coupling between the windings.

Transformer Failure

Typical simple transformers are found in functions such as audio output, vertical output, etc. They typically open in the

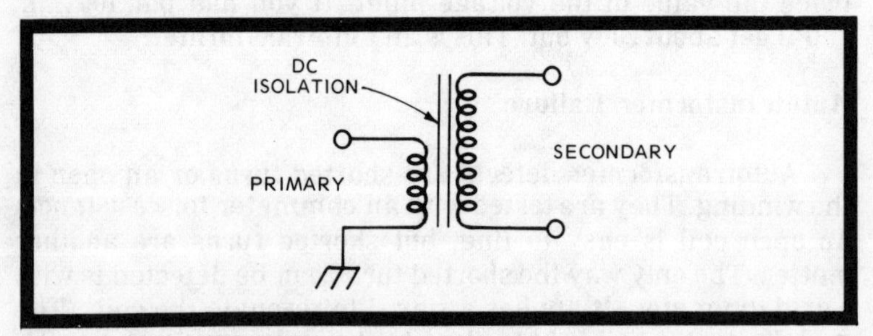

Fig. 17. An isolation transformer has good ac coupling, but dc cannot get from primary to secondary.

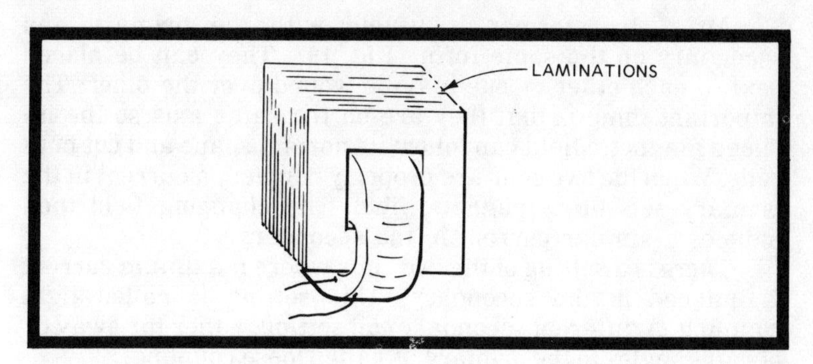

Fig. 18. In order to reduce eddy current losses, a transformer iron core is made of laminated layers.

secondary since there are generally large amounts of dc voltage present. The open is easily detected with an ohmmeter.

Transformer windings do not short too often, but do develop another unusual problem. The iron cores are laminated (Fig. 18) or are composed of a number of thin insulated strips. The varnish insulation ages and the laminations loosen, which causes distortion of the function.

TYPICAL RF TRANSFORMER 10

In a power transformer or choke, a frequency of 60 Hz or thereabouts is used. In order to attain large values of inductance, an iron core is used to concentrate the magnetic flux. In an rf transformer, frequencies are in the kilohertz or megahertz range. At these frequencies ordinary iron cannot be used as a core because high-frequency current builds up heat. This heat loss is called an eddy current loss. Also, the iron resists rapid rf changes in the magnetic field. As a result, hysteresis losses take place.

Rf transformers need a special iron core, but the iron is ground into tiny pieces and mixed into a hardened material with insulation. That way the magnetic field passes through mostly insulation with bits of iron here and there. This lowers the permeability of the iron core and reduces the amount of inductance in the transformer. An rf transformer need have only small values of inductance; typical values are in millihenrys and microhenrys. Between the powdered-iron core and the tiny amounts of inductance, moving the core causes large-percentage variations of inductance.

An rf transformer is wound with the primary and secondary on the same form (Fig. 19). They can be placed next to each other or one may be wound over the other. The important thing is that they are on the same axis so the induced magnetic field can follow its normal shape and cut both coils. When the two coils are properly coupled, a current in the primary sets up a magnetic field. The changing field then induces a similar current in the secondary.

There is a setting of the two coils where maximum current is induced in the secondary. This setting is called tight coupling. A different secondary coil setting, either far away or at right angles to the primary, is called loose coupling.

In an rf transformer, the primary and secondary coils must be "tuned" (Fig. 20). Should one or both of the coils be detuned, the rf signal will not be coupled properly from the primary to the secondary. Depending on the amount of detuning, the signal could be distorted or even lost altogether.

Usually, the coils are tuned with a shunt capacitor and the tuning core. The capacitance is fixed and the core can vary the coil inductance over a wide range. The tuned frequency of the circuit is determined by the LC combination. Transformers are carefully wound on forms that have a specific diameter, with a certain number of turns per inch. This determines the inductance of the coil.

RF Transformer Failure

An open in an rf transformer is located quickly with an ohmmeter. When a few turns short, the isolation is not so easy.

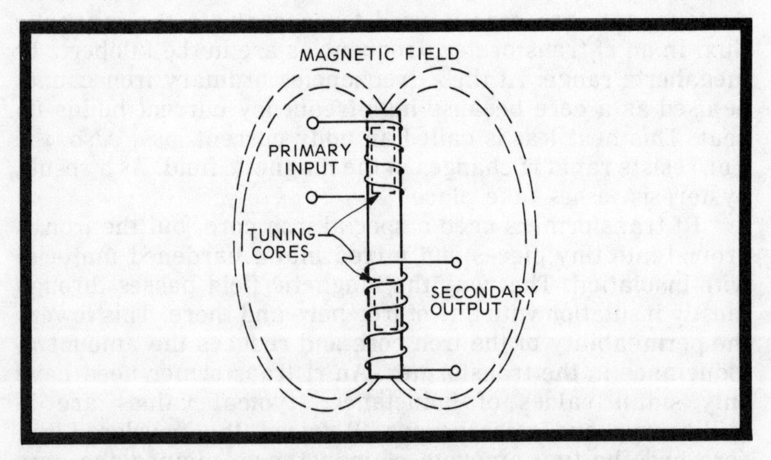

Fig. 19. Rf-type transformers need only a few turns in the windings due to the low current and high frequencies involved.

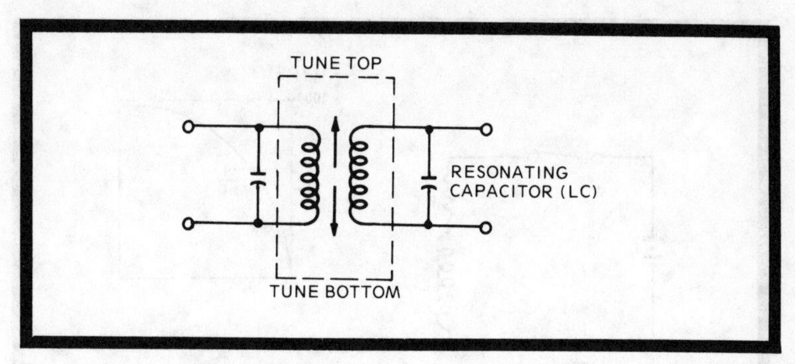

Fig. 20. An rf transformer operates at best efficiency when it is tuned to the frequency being worked.

Resistance readings prove little.

Aside from direct substitution, the best way to test the transformer is to try to align it according to factory alignment specifications. If during the alignment it is found that the waveshapes that are supposed to be present are distorted or absent, the transformer is a prime suspect. It is a good idea to make sure it is bad before going into desoldering.

LR TIME-CONSTANT CIRCUIT (FLYBACK)

11

The LR time-constant circuit is something like the RC time constant and a lot of the mathematics look alike. However, since the servicer is rather more interested in application than mathematics, let's examine a flyback transformer used in a TV high-voltage system.

Of course, an LR time constant is an inductance and resistance in series (Fig. 21). The time constant is the time in seconds it takes for the circuit to rise to 63 percent of full current, which is determined by dividing the resistances into the inductance. The inductance is in henrys and the resistance in ohms.

If you have a battery, switch, resistance, and inductance in series, the charging time begins when the switch is closed. In the coil, a magnetic field is induced. The field then induces a back voltage that tries to slow the electron flow. Between the resistor and the back voltage, the electrons are slowed and the charge reaches 63 percent at the time constant value.

In a flyback transformer this effect is taken one step further to produce the high voltage. There is no actual resistor per se in the flyback. The resistance is that of the coil itself.

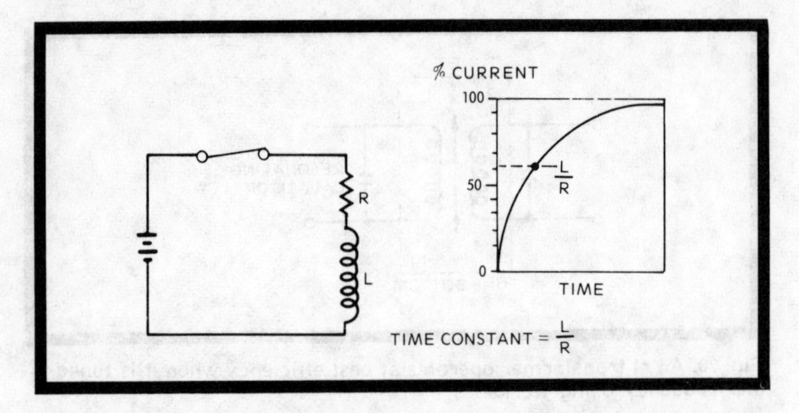

Fig. 21. The time constant of an inductance and resistance in series is the interval it takes for the circuit current to reach 63 percent of full current.

The flyback coil is fed a sawtooth voltage from the horizontal output section. A sawtooth wave has a slowly rising voltage that reaches its peak, then suddenly returns to minimum voltage.

As the slowly rising voltage passes through the flyback coil, the gradually rising voltage induces a rising magnetic field, which induces a back voltage. The back voltage is considerable but not a high voltage. The back voltage induced is proportional to the speed of the developing magnetic field. The operating frequency, 15.75 kHz, is fairly high, and a significant back voltage is induced.

Then the sawtooth reaches its peak. It tops out and the voltage drops to minimum. The speed of the voltage collapse is high. The magnetic field drops almost instantaneously from a large value to zero. This induces an abnormally high back voltage. In a color TV it can reach 25 kV. The voltage is there for only an instant, but it is fed into a charged high-voltage filter capacitor which maintains its charge. The high voltage is then passed onto the crt to do its job.

Flyback Failure

Flyback transformers work at such high voltage values that the typical indication of breakdown is visible: they burn or char. On occasion, though, they develop shorted turns or open up. A point-by-point resistance check picks out open windings, but shorted turns cannot be ordinarily located in that fashion. Special flyback grid-dip meters have to be used to determine the existence of shorted turns. Since they are only about 90 percent accurate, you can but cross your fingers and solder in a new replacement.

IMPEDANCE-MATCHING CIRCUIT 12

Electronic circuits have output resistance and input resistance. If the two resistances are the same, the output from one circuit can be attached directly to the input of another and the electrons will travel without significant loss. If the two resistances are different, electrons will be reflected at the junction and bounce back in the direction they are coming from, causing losses and distortion.

The resistance is called "impedance," denoted by the letter Z instead of R, since impedance is a combination of resistance and reactance. Reactance, like resistance, is measured in ohms and is caused by the effects of capacity and inductance on voltage and current. (See **Reactance Circuits** for further explanations.) The servicer would do well to consider reactance as a form of resistance.

A typical audio output tube (Fig. 22) has an output resistance of about 5000 ohms and an ordinary loudspeaker has an input resistance of 4 ohms. (Remember, these resistances are a combination of reactance and resistance.) So, an audio output transformer is used to match the two circuits.

A large primary inductance, to provide the tube with a 5000-ohm load, comprises many turns of wire. A small secondary inductance, to provide the speaker with a 4-ohm input, is made up of a few turns of wire.

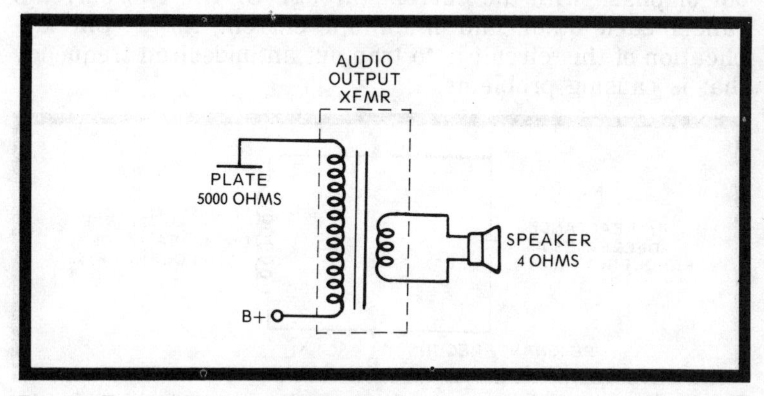

Fig. 22. The most important function of an audio output transformer is matching the amplifier to the speaker.

Output Transformer Failure

Audio output transformers short, open, and sometimes leak through the insulation of the windings. On occasion, the laminations loosen and vibrate.

13 SIMPLE TANK CIRCUIT

If you put a capacitor and coil in parallel with each other, they form what is known as a tank circuit, a trap for a particular frequency (Fig. 23). At this frequency, a minimum of current passes through. At all other frequencies, higher amounts of current flow.

A capacitor has a high reactance at low frequencies. That's because a lot of electrons flow into the capacitor as the electron flow changes direction at the slow pace. The same capacitor has less and less reactance as the frequency is raised. That's because fewer electrons are forced into the capacitor as the ac reversal picks up speed. Therefore, as the frequency is increased capacitive reactance drops.

On the other hand, a coil has very low reactance at low frequencies. That's because the back voltage is small at low frequencies, but increases as the frequency goes up. The larger the back voltage, the more the voltage drop due to reactance.

Since C and L develop opposite amounts of reactance as the frequency changes, there is some particular frequency where the two reactances are the same. This is the **resonant** frequency. At that frequency, the current through L is exactly out of phase with the current through C. The two currents cancel each other and minimum current flows. One application of this circuit is to trap out an undesired frequency that is causing problems.

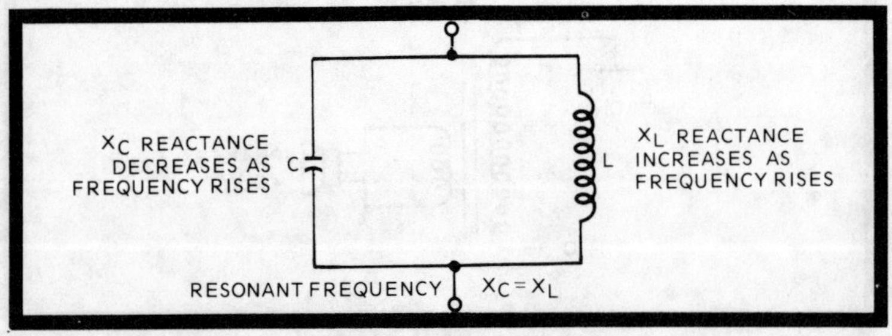

Fig. 23. The resonant frequency of a tank circuit is determined by the values of inductance and capacitance in shunt.

PEAKING COIL CIRCUIT 14

As you increase the frequency of a signal in a coil, the back voltage developed causes a higher and higher reactance. The higher the reactance, the more the voltage drop across the coil (Fig. 24). This is a valuable characteristic in many applications. For instance, in a video amplifier frequencies of 0-3.5 MHz have to be amplified. Ordinary amplifiers lose gain as frequency is increased and can't usually amplify anything above a megahertz. A peaking coil in series with the output can extend the frequency response.

As the frequency being amplified rises, the reactance in the peaking coil increases. It's just as if the resistance in the output increases as the frequency rises. The amplifier provides less gain at the higher frequencies, but the varying reactance compensates by developing larger amounts of signal across the increasing reactance.

Peaking Coil Failure

In a video amplifier, when the frequency response falls off, the TV picture "smears." The smeared picture is usually due to an open video peaking coil. The test is easy—short the peaking coil. If the smear disappears, the coil is open. (It usually opens on one end.) A drop of solder can cure it, or a new one should be installed. While it is possible for a peaking coil to short, it is extremely rare.

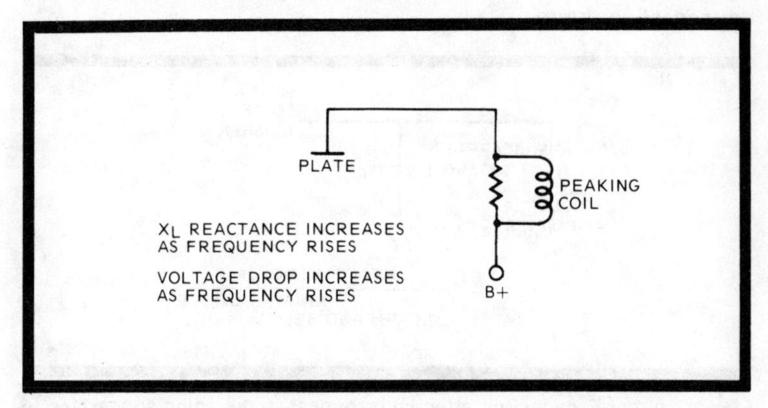

Fig. 24. A peaking coil is typically used in a video amplifier to improve the output of high frequencies.

15 BYPASS CAPACITOR CIRCUIT

Bypass capacitors are common in all electronic gear. A typical bypass application is found in color amplifier output circuits (Fig. 25). In the output of a color amplifier there may be a component of the 3.58 MHz color subcarrier along with color intelligence at frequencies up to 1.5 MHz. The color subcarrier must be eliminated or it will appear in the TV display.

A bypass capacitor of about 39 pF is installed in the amplifier output. As the signal passes through the output line, it arrives at the 39 pF capacitor, and the 3.58 MHz signal is bypassed to ground through the capacitor while the color signal keeps right on going to the crt.

A 39 pF capacitor has very little reactance at 3.58 MHz. Below 1.5 MHz, the bypass has a very high reactance, so the capacitor acts as an open circuit. The 1.5 MHz signal is unaffected as the 3.58 MHz component is bypassed to ground.

Bypass Failure

Since the bypass is in the output, there is usually some dc voltage on it. This can cause the capacitor to short, open, or leak. A shorted capacitor lets all the signal and dc voltage go to ground, causing resistors and other current-carrying components to overheat. An open capacitor doesn't do any bypassing, so the 3.58 MHz signal distorts the TV display. A leaky capacitor can act shorted or open, according to the degree of leakage.

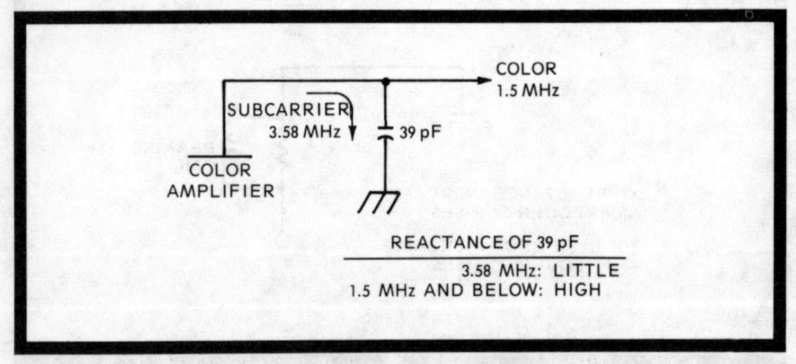

Fig. 25. A 39 pF capacitor effectively bypasses the color subcarrier to ground.

LIGHT BULB CIRCUIT

16

The common light bulb is an electronic device you can call a monode. It is a vacuum tube with one element, a filament or cathode. When ac or dc is passed through the filament, the electrons heat the filament. Those near the surface free themselves from the wire and drift into the vacuum. They form a "space cloud" of ions around the filament. The space cloud, being ionic, has a negative charge, repelling further electron "boiloff." The space charge attains a certain negative voltage and maintains it (Fig. 26).

If you increase the power source voltage, more electrons are forced into the space cloud, increasing the negative charge. If you decrease the power source, the negative charge drops. All this is inconsequential to the function of the light bulb, but it becomes vital if you add another element to the light bulb, as is shown in the next circuit.

Light Bulb Failure

A light bulb filament, of course, opens and the bulb must be replaced. Also, the filament resistance can increase from cracking. This can cause sparking in the bulb, which makes the simple bulb an rf transmitter that can cause all kinds of interference.

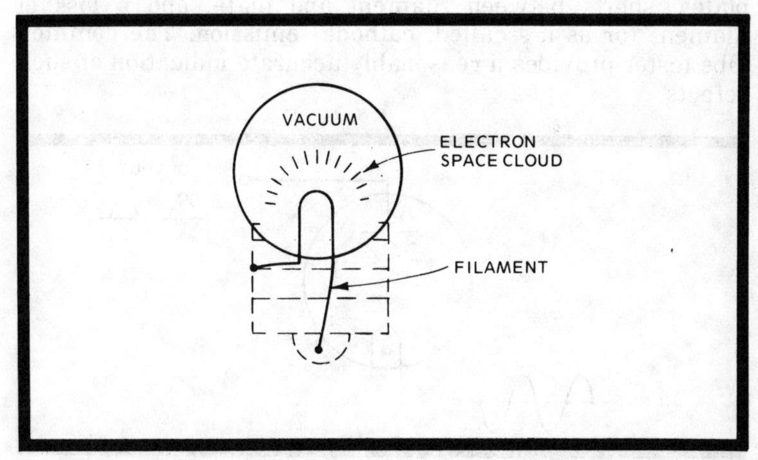

Fig. 26. A light bulb is the basic building block of a tube. I call it a monode. It only has a heater or filament.

17 VACUUM TUBE DIODE CIRCUIT

If you add another element to a light bulb and put a positive charge on the second element with a battery, electrons will leave the electron-rich space cloud and go toward the electron-poor positive element (Fig. 27). If you keep the positive charge on the plate, a continuous flow of electrons takes place. Even if the filament voltage is ac, the electrons may flow in but one direction in the tube.

Rectification

Since electrons can flow only in one direction, the diode can be used to change ac into dc. Simply put ac voltage across the diode instead of a battery. During the positive cycle, the plate has a positive charge and the filament a negative charge. Electrons will flow. During the negative cycle, the plate gets a negative charge and the filament a positive charge. The plate repels any electrons that try to go there. Therefore, electrons flow only while the positive cycle is present. The ac is now dc.

Diode Failure

Typical diode tube defects are open filaments, open plates, shorts between filament and plate, and a loss of filament (or as it's called, cathode) emission. The common tube tester provides a reasonably accurate indication of such defects.

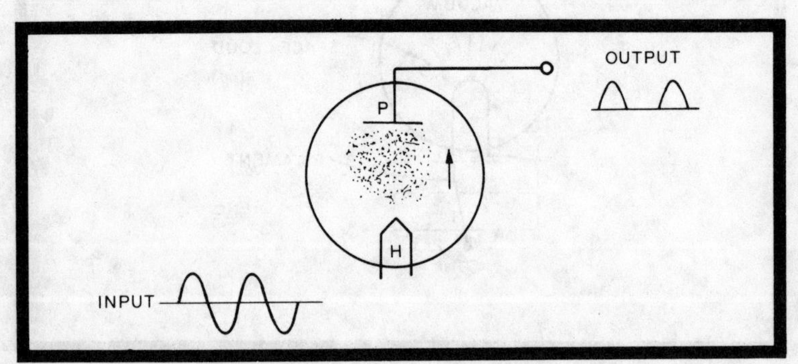

Fig. 27. A diode tube has a heater and a plate. Electrons can flow from H to P but not the other way.

SOLID-STATE DIODE CIRCUIT 18

Solid-state diodes have been around as long as vacuum tubes. The original cat-whisker crystal (Fig. 28) radio used a solid-state diode. The modern solid-state diode is made up of two pieces of semiconductor material, typically silicon or germanium crystals, doped to produce so called n material and p material. N material is silicon doped with arsenic or antimony. This gives the inert crystals extra electrons. The extra electrons can move freely through the crystal structure. The electron freedom of movement changes the normal silicon nonconductor to a semiconductor. It's called n material because the electrons represent a negative electron path for incoming electrons from a power source. The n material provides a resistance-free path, just as the vacuum in a vacuum tube allows electrons to move unimpeded.

If the silicon is doped with aluminum or gallium, the crystal becomes a p-material semiconductor. The doping causes a deficiency of electrons in the crystal structure. The resulting holes are positive charges. The silicon is then called p material because the lack of electrons represents a positive electron path for incoming electrons from a power source (Fig. 29).

There are no negative or positive charges on the n or p material. The negative or positive pathways are simply means of changing normal nonconductive crystals to semiconductors in which the resistance to electron flow has been reduced.

The pieces of n and p semiconductors by themselves are not too valuable. But when the two are joined together, they form a junction which becomes a diode (Fig. 30). The p material becomes the anode side, like a tube's plate, and the n material becomes the cathode like a tube's filament.

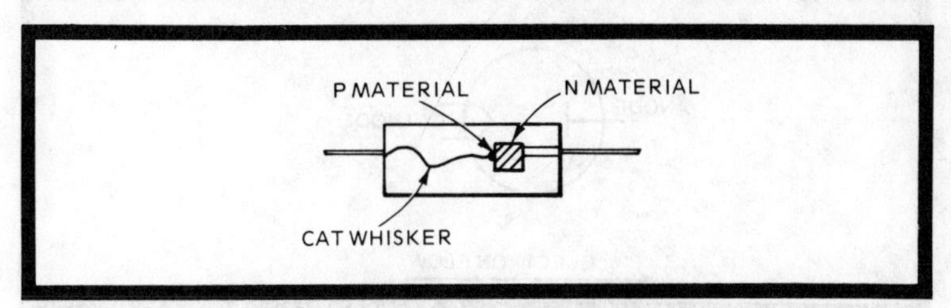

FIG. 28. A cat-whisker diode is made of a piece of n material and a stiff piece of wire.

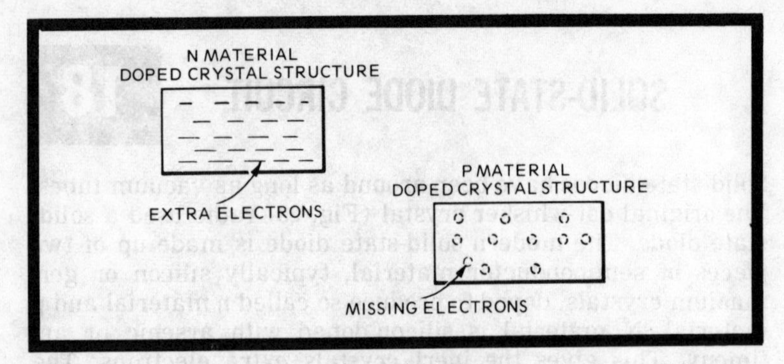

Fig. 29. N material has an excess of electrons. P material has deficiency of electrons.

However, electrons can travel from cathode to anode in the semiconductor diode without the need for a hot filament.

When the junction is formed, excess electrons in the n material suddenly are confronted with holes in the p material (Fig. 31). The two charges are attracted to each other. At first, some electrons in the front line jump across the junction into holes. That neutralizes the junction. Then a few more electrons jump into the neutralized structure. This places an excess of electrons on the p side of the junction and a

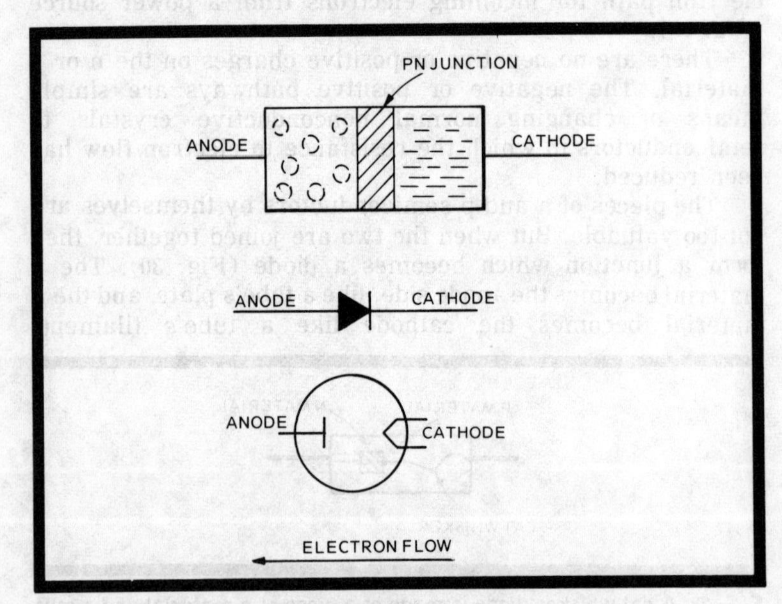

Fig. 30. Electrons flow from the cathode to the anode in a diode.

deficiency of electrons on the n side. This is opposite to their status quo. A slight negative charge develops on the p side and a slight positive charge develops on the n side. That is a good pn junction. It has a slight voltage force of about 0.1V. This charged area is the basis for all diode and transistor action. The servicer would do well to visualize the junction charge as he examines a solid-state circuit.

Junction Bias

While the pn junction maintains this tiny voltage without outside influence, manipulation of this bias voltage provides diode action. If the junction bias voltage is made bigger, the physical size of the junction will spread. If the bias is made smaller the junction will practically disappear. The larger the junction, the larger the resistance will be to electron flow. This is called reverse bias. The smaller the junction, the smaller the resistance. This is called forward bias.

Reverse bias is attained by attaching a dc power source with the positive to the cathode and the negative to the anode. That way, the positive charge on the cathode n material attracts the electrons away from the junction, tending to increase its size. The negative on the anode p material attracts the holes away from the junction, also increasing the junction resistance.

Forward bias is attained by attaching a dc power source with the positive to the anode and the negative to the cathode. That way, electrons flow to the junction and close the width of the junction. At the same time the positive terminal attracts the electrons, thus reinforcing the closing of the barrier. Under this condition there is a continuous flow of electrons through the diode.

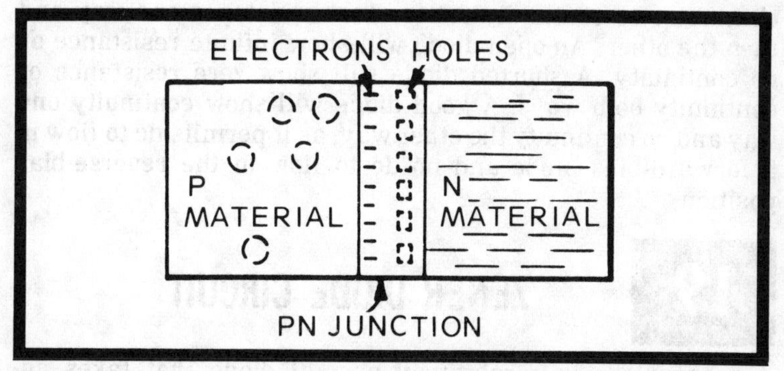

Fig. 31. A pn junction in a diode forms a barrier to the electron flow from n to p.

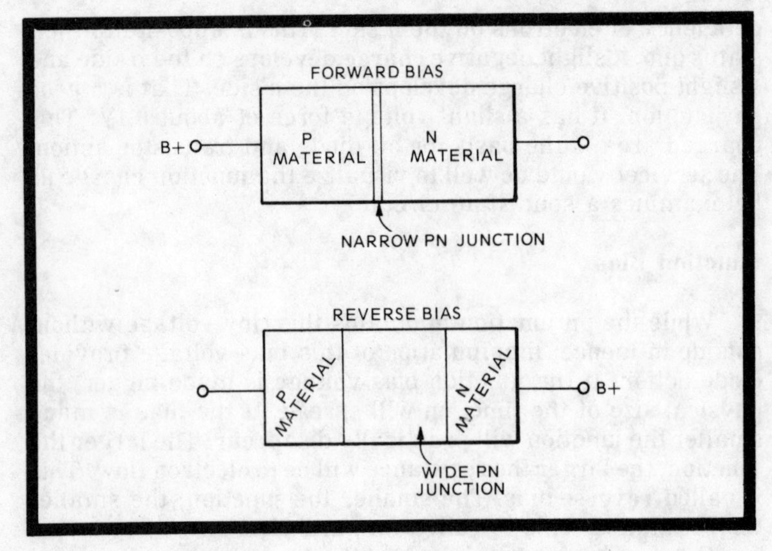

Fig. 32. The pn junction barrier changes in width with the amounts of voltage applied.

If an ac voltage is applied to the diode, a current will flow when the diode is forward biased and stop when the diode is reverse biased. This is changing ac to dc, since electrons flow only in one direction.

Diode Failure

A solid-state diode fails by either opening or shorting. It can probably happen, but I've never seen one simply leak. When the junction ruptures, the diode shorts completely.

Testing a diode is simple and accomplished with either an ohmmeter or continuity tester. Test resistance one way and then the other. An open diode will show infinite resistance or no continuity. A shorted diode will show zero resistance or continuity both ways. A good diode will show continuity one way and no continuity the other way, as it permits dc to flow in a forward-bias mode and no dc to flow in the reverse-bias position.

19 ZENER DIODE CIRCUIT

A zener diode is a relatively normal diode that takes advantage of a normally unused diode characteristic. Diodes are typically used in rectification and are operated either in a forward-or reverse-bias mode. A zener diode is operated deep in the reverse bias range, way past normal reverse-bias settings (Fig. 33). At a deep reverse bias the diode suddenly loses all its semiconductive ability and becomes a dead short. Then, no matter how many volts are applied, the current flow is so heavy that the voltage at the diode cannot rise above that reverse-bias setting, the point where it turned into a dead short.

Zeners are designed to become a dead short at a particular point. If a zener is lightly doped, the zener "knee" voltage (so named because the graph displays a knee, Fig. 34) will be high, in the 40V or better range. If the zener is doped with medium amounts of arsenic in the n material and aluminum in the p material, the knee voltage can be around 20V. Should the zener be heavily doped, its reverse breakdown characteristic can be near 9V.

Zener diodes are used in power supplies to hold a particular voltage at a steady state no matter what the incoming voltage might do. The cathode of the diode is attached to the positive voltage point in the power supply. That way, it has a reverse voltage on it. No conduction can take place in a normal way. For instance, if it is a 9V zener and is placed into a supply where there is some variation above 9V, any voltage variation simply increases or decreases the electron flow from anode to cathode (reverse electron flow), thus holding the voltage securely at 9V.

The zener has a power rating. The rating is in watts just like a resistor. The zener diode is actually a resistor of very low ohmage to any voltage above its knee voltage. Below the knee voltage it is an ordinary diode reverse biased.

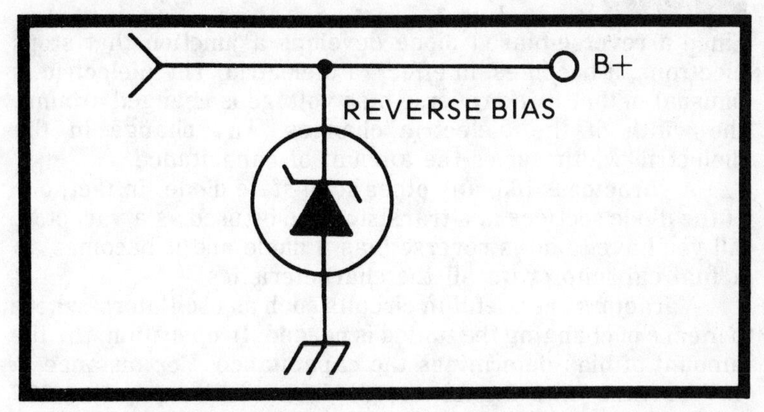

Fig. 33. A zener diode is typically operated with a reverse bias.

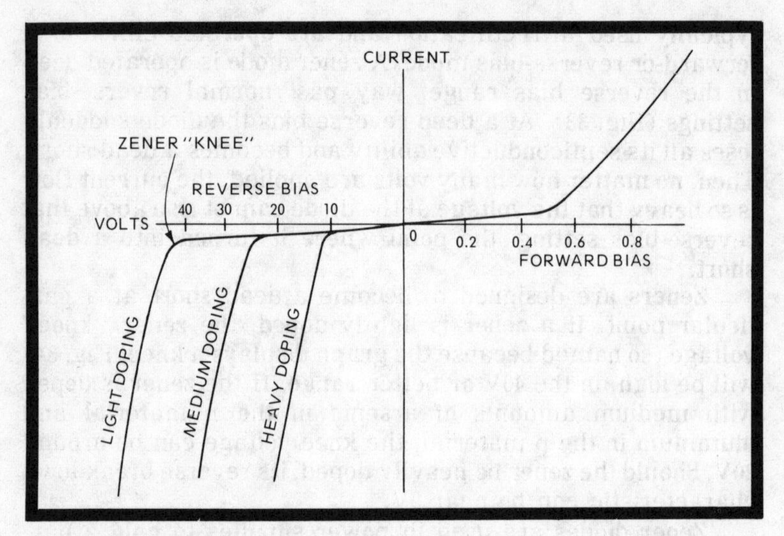

Fig. 34. The knee voltage point determines the amount of voltage a zener is designed to regulate.

Zener Failure

Zener diodes fail in the same way that ordinary diodes do and are tested in exactly the same way. The junction can rupture and cause a short; but, more likely than not, they open. Then, the voltage regulation ceases and too much voltage gets into the circuit.

20 VARACTOR CIRCUIT

Since a reverse-biased diode develops a junction that stops electrons, it becomes, in effect, a dielectric. The dielectric is unusual in that if the reverse-bias voltage is changed around, the width of the dielectric changes. Any change in the dielectric width varies the amount of capacitance.

A varactor is like any other solid-state diode. In fact, one of the diode sections in a transistor can be used as a varactor. All you have to do is reverse bias a diode and it becomes an actual capacitor with all the characteristics.

Varactors are useful in circuits such as oscillators, where a means of changing the tuning is needed. In operation, the the amount of bias determines the capacitance. For instance, a 10V reverse bias makes the diode a 4 pF capacitor; 20V changes it to 6 pF, 30V makes it 8 pF; 40V, 10 pF; and so on.

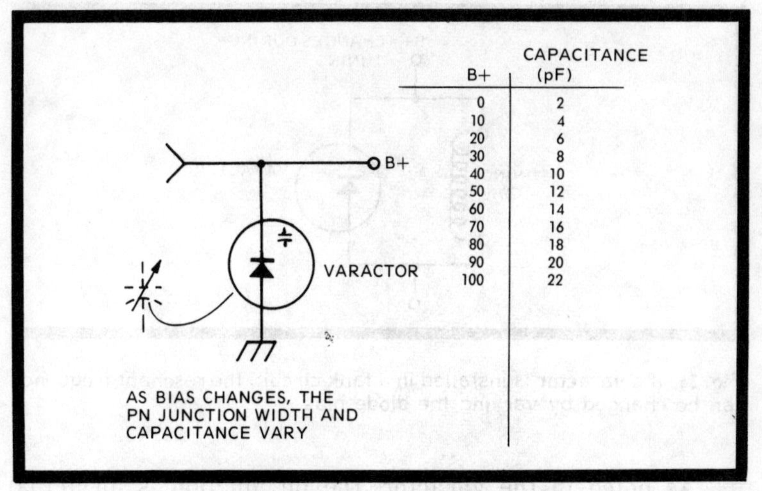

Fig. 35. A varactor diode must be reverse biased in order to function as a variable capacitance.

About a 10V change gives the varactor about 2 pF more or less (Fig. 35).

In some new color TVs, instead of using the normal switch or turret tuner that changes frequency by physically changing a set of coils (inductance), varactors are used. By simply pressing a button, the dc reverse bias is changed and the frequency is changed as the capacitance changes.

Varactor Failure

Varactor diodes must have a junction that has very little reverse leakage. Any leakage at all puts resistance in series with the capacitance and lowers the capacitance effect due to the time constant formed. The more the leakage the less effective the varactor.

A varactor can, of course, short or open, but is is usually used in a circuit with no real power involved. Just make sure that when you test a varactor it has little or no leakage.

CAT-WHISKER DIODE CIRCUIT 21

A germanium diode used in detector applications has a pn junction but no p material. The germanium crystal is doped with arsenic and becomes a piece of n material. Then when it comes time to weld it to p material a problem arises.

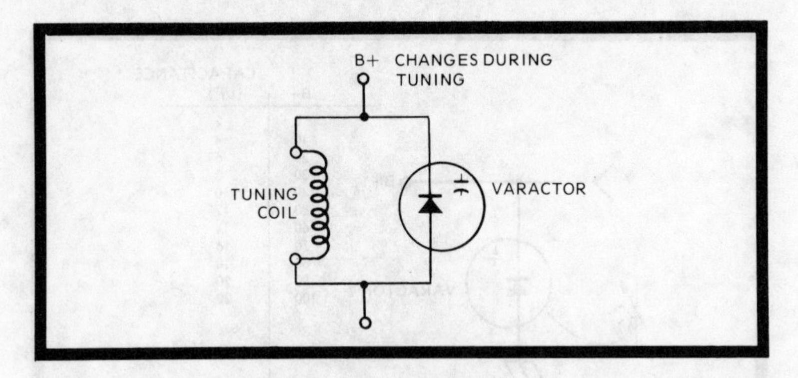

Fig. 36. If a varactor is installed in a tank circuit, the resonant frequency can be changed by varying the diode bias.

As noted in the varactor, the pn junction is an actual dielectric when reverse biased. The larger the surface area, the more electrons can pile up on the negative side and the more the electrons on the positive side are repelled. This large storage represents large capacitance.

It is desirable in a detector-type diode to have little or no capacitance. For if you try to detect rf or i-f in the kilohertz or megahertz range, as the frequency goes up so do the losses. To circumvent the frequency losses, the surface area of the junction is made as small as possible. A piece of stiff wire (cat-whisker) is attached to the n material (Fig. 28). During manufacture, as the wire is attached a small amount of p material is produced at the junction. The tiny point contact reduces capacitance drastically. The low-capacitance diode is thus formed.

Germanium Diode Failure

A germanium diode can short, open, or develop junction leakage. An ohmmeter will detect a bad one.

22 TRIODE VACUUM TUBE

If we introduce a third element between the anode and cathode in a vacuum diode, we have a triode. The third element, called a control grid, can vary the flow of electrons in the tube (Fig. 37). The control grid actually acts like a variable resistance or a valve (Fig. 38) to electron movement.

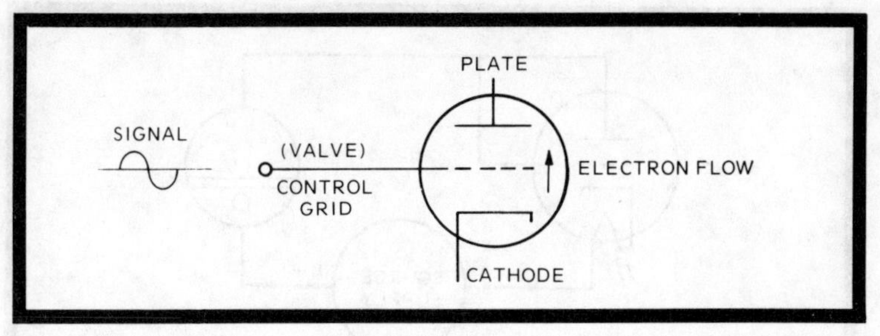

Fig. 37. A triode has a control grid between the cathode and plate. The grid regulates the flow of electrons.

In a diode, a space charge of negatively charged electrons forms around the cathode. A positively charged plate exerts an attraction to the electrons. The electrons start a journey to the plate. In a triode, the electrons encounter the wire mesh grid just past the cathode. The grid has an electrostatic charge on it. If the charge is a high negative voltage the electrons are repelled back to the cathode space charge. If the charge is positive, the electrons are attracted to the grid and zoom through the mesh, heading for the plate. Should the charge be an in-between voltage, some of the electrons can pass through the mesh according to the voltage level. If the charge is positive, not all of the electrons get through the grid. Some stop at the grid and flow into the grid circuit. This is control grid current and constitutes a loss of electrons that never reach the plate.

Since the control grid is so close to the cathode, it exerts a much larger electrostatic influence than the plate. If a sine-wave signal is fed into the control grid, the voltage level on the control grid varies. The varying control grid voltage causes a

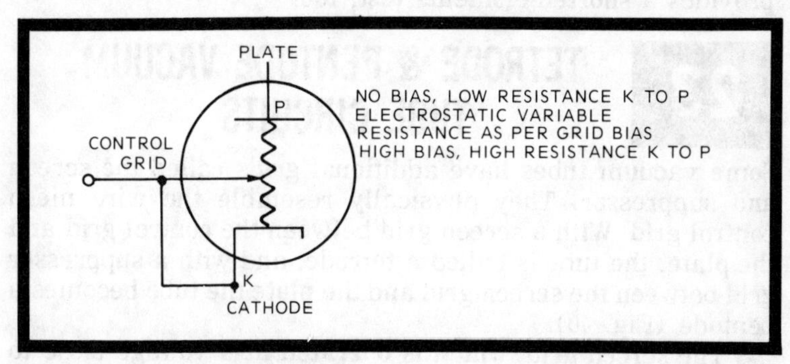

Fig. 38. The control grid acts like a resistance to the electron flow. The resistance varies with grid bias.

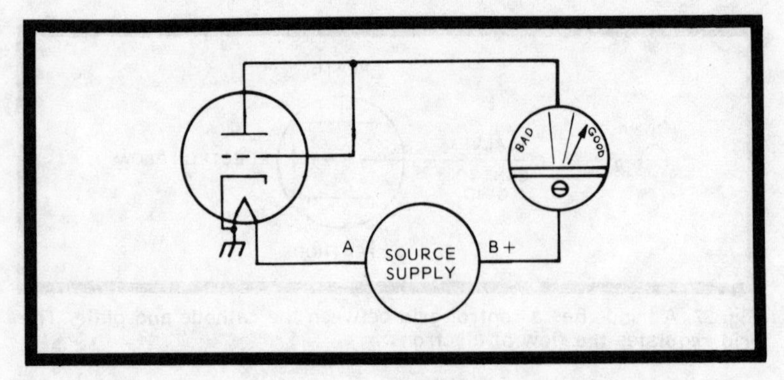

Fig. 39. The basic tube test is a measure of cathode emission. It is assumed the other characteristics will vary as the emission does.

similar variation in the intensity of the electron flow from cathode to plate. If the grid voltage variation is 2V and it varies the plate voltage by 20V, the change in the plate is ten times greater than the grid voltage change. Assuming the plate change looks exactly like the grid sine wave, the triode will have amplified the sine wave.

Triode Failure

Most triodes can be checked with the common tube tester. While factory-type testers provide more sophisticated checks, a triode is usually tested like a diode (Fig. 39). The amount of cathode emission is measured by connecting the control grid to the plate during the test. That way, there is no valve action and the same positive voltage on the plate and grid allows the space charge to be attracted to the plate. The test button should not be held down too long, since during the test the tube is conducting heavier than it is designed to. The tester usually provides a shorted-elements test, too.

23 TETRODE & PENTODE VACUUM TUBE CIRCUITS

Some vacuum tubes have additional grids called the screen and suppressor. They physically resemble the wire mesh control grid. With a screen grid between the control grid and the plate, the tube is called a tetrode, and with a suppressor grid between the screen grid and the plate the tube becomes a pentode (Fig. 40).

The screen grid, which is operated at a voltage close to that on the plate, has the effect of giving the electrons additional acceleration toward the plate. The suppressor grid is given a cathode-level voltage, but by the time electrons reach the suppressor they are traveling so fast they go through the suppressor despite the low voltage.

As electrons hit the plate, some of them bounce off or dislodge other electrons from the plate. Without the suppressor, which forces them back to the plate, these free electrons (called secondary emission) would flow back to the positively charged screen grid, become screen grid current and thereby be lost.

There is another vital reason for the screen grid. This has to do with unwanted capacitances between the cathode, control grid, and plate electrodes of the triode (Fig. 41). These capacitances are called interelectrode capacitance. During high-frequency applications, interelectrode capacitances act as coupling capacitors and feed signal back from the plate to the control grid and cathode. The resulting distortion and losses can be great. By adding the screen and a bypass capacitor, the plate-to-grid coupling is eliminated.

The input resistance of a tube is the resistance introduced by the grid bias to the signal as it modulates the electron stream. The output resistance is the plate resistance of the tube. Actually, it is not the plate resistance itself, but the resistance in parallel with the output capacitance, just as the input resistance has to contend with the interelectrode capacitances, too.

What is the plate resistance? For a given grid bias voltage, changes in plate voltage will cause more or less signal

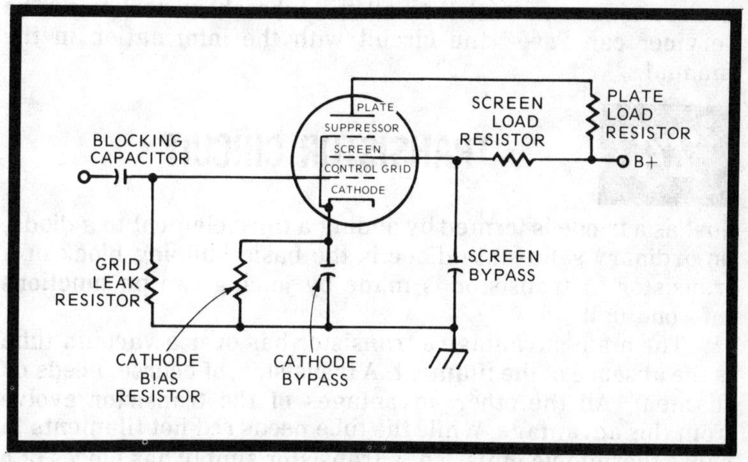

Fig. 40. A typical pentode amplifier circuit requires the components shown.

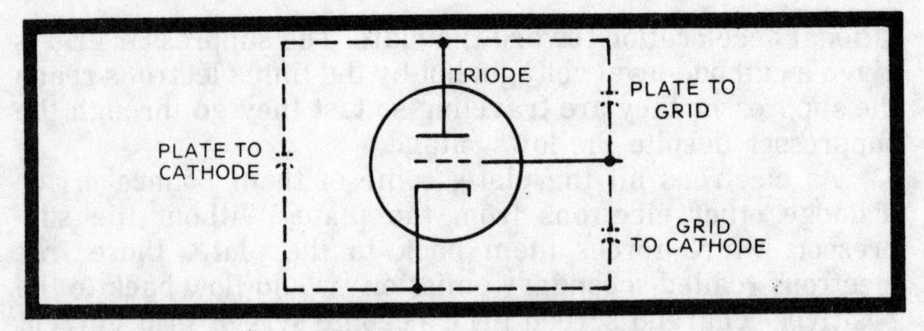

Fig. 41. There are a number of invisible interelectrode capacitances inside a tube.

current to be produced in the plate circuit. If the plate voltage rises, more signal current will be produced as the stronger positive voltage attracts more cathode-to-plate current. Should the plate voltage drop off, less signal current will be produced. The plate resistance is the plate voltage change divided by the change in plate current caused by the voltage variation. Plate resistance is measured in ohms; R equals E (plate voltage change) divided by I (plate current change). A typical tube could have an output resistance of 10K if 1V of plate voltage change causes 0.1 mA (1 divided by 0.1 equals 10K).

Tetrode & Triode Failure

A servicer should know about input and output resistances and interelectrode capacitances to intelligently service tube circuits. Many times a schematic of a particular unit is not available and a servicer can use a tube manual to troubleshoot a circuit. The manual lists all the tube characteristics. The servicer can "see" the circuit with the information in the manual.

24 TRANSISTOR CIRCUIT

Just as a triode is formed by adding a third element to a diode, an ordinary solid-state diode is the basic building block of a transistor. A transistor is made by joining two pn junctions into one unit.

The main advantage a transistor has over a vacuum tube is the absence of the filament. A transistor, of course, needs no filament. All the other advantages of the transistor evolve from this advantage. While the tube needs red-hot filaments to cause thermionic emission, a transistor simply has pieces of n and p material that afford the electrons passage.

A pnp transistor consists of a single slab of neutral (intrinsic) material. The two ends of the slab are doped with a p material, and the center section is doped with an n material. An npn transistor, also a single slab of an intrinsic material such as silicon or germanium, is doped with n material on both ends and p material in the center. Either type of transistor might be considered as two diodes sharing one element.

If you think of a transistor as back-to-back diodes, testing them becomes easy. An ohmmeter or continuity tester gives an accurate go, no-go test. Each diode section can be tested for an open, short, or leak. (A diode must read a high resistance in one direction and a low resistance in the other. Any other readings, such as high in both directions, low in both directions, or variations indicate a defective transistor.)

One of the confusing things about transistors, in comparison to vacuum tubes, is that electrons flow from emitter to collector or collector to emitter. It is said that an emitter is like a cathode and a collector like a plate. How can the collector send electrons to the emitter? An npn has a cathode of a diode as the emitter. When the emitter is a cathode, electrons flow from emitter to collector. On the other hand, a pnp has the anode of a diode as an emitter. When the emitter is an anode, electrons flow from collector to the emitter (Fig. 43). Electrons flow to the anode from the cathode, whichever the emitter is.

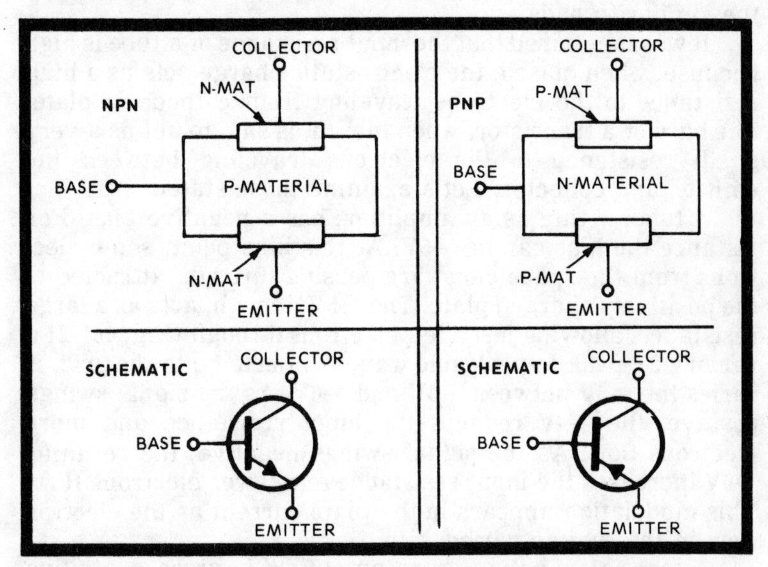

Fig. 42. There are two types of bipolar transistors. Each has its own specific uses.

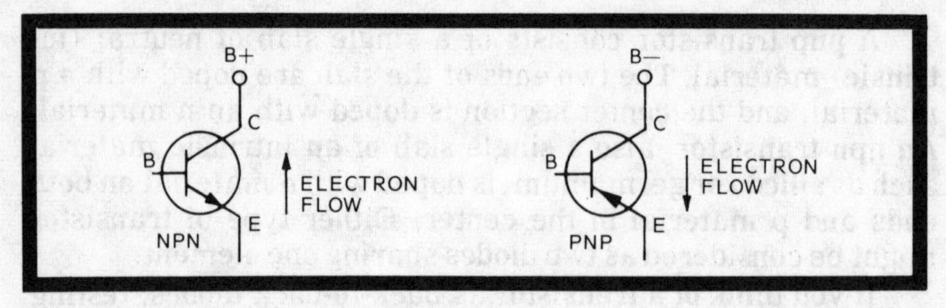

Fig. 43. With an npn transistor B-plus is applied to the collector. B-minus goes to the collector of a pnp.

That is why a transistor can have either B-plus or B-minus on the collector. B-plus voltage is used to supply npn transistors and electrons flow from emitter to collector, attracted by the positive voltage or the deficiency of electrons. B-minus voltage is used to power pnp transistors. With B-minus on the collector and almost ground zero on the emitter (the emitter is actually P material), excess electrons leave B-minus, pass through the transistor from collector to emitter and then to ground.

Bias

It is said that the base of a transistor is like the triode control grid because the signal is applied to the base. But there the similarity ends.

It was mentioned that the input resistance of a tube is high because, when biased, the electrostatic charge acts as a high resistance to the electrons traveling from cathode to plate. The base of a transistor, when biased, is said to act as a very small resistance to the electrons traveling between the emitter and collector. Let's examine those statements.

A tube, acting as an amplifier, has a negative bias. For instance the bias can be –5V. At this bias point, some electrons from the space cloud are passing through, attracted to the positively charged plate. The –5V, though, acts as a large resistance, allowing just a few electrons through (Fig. 44). If a varying 2V peak-to-peak sine wave is injected into the grid, it varies the –5V between –4V and –6V. As the signal swings positive, the –4V reduces the input resistance and more electrons flow. As the signal swings negative, the resulting —6V increases the input resistance and fewer electrons flow. This modulation appears in the plate current as the electron flow in the tube is varied.

A transistor, acting as an amplifier, is given a positive forward bias. For instance, a silicon npn transistor is given a

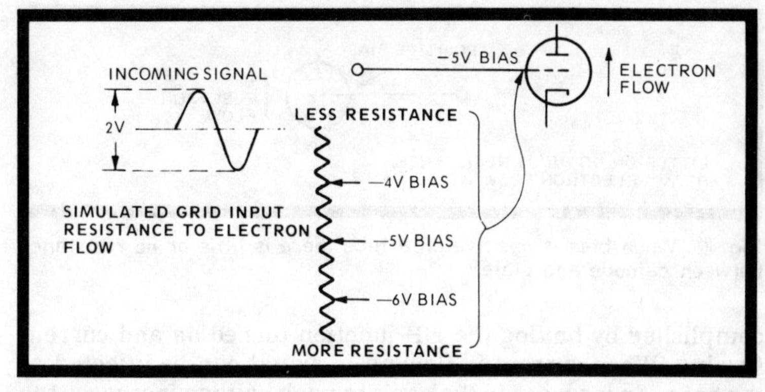

Fig. 44. The bias on a tube amounts to an electrostatic resistance to the electrons.

0.6V positive voltage on the base in relation to the emitter; the 0.6V turns on the transistor, since the pn base-emitter junction is forward biased (Fig. 45). The pn junction width is practically gone with the forward bias. Electrons flow freely from the emitter to the positively charged base. Electrons also flow with the same intensity from the emitter to collector. The base provides little or no resistance to the emitter-collector electron flow. That is why it is stated that tubes have a high input resistance and transistors a low input resistance.

If you'd like, you can set up a tube to have a low input resistance. All you have to do is give the control grid a positive voltage (Fig. 46). This makes the grid exercise no resistance to the electrons. They flow as if the grid wasn't there. In the same manner, you can make a transistor have a high resistance (Fig. 47). Simply put a large reverse bias on the EB junction. The reverse bias stops the EB flow and also becomes a large resistance to the EC flow.

The preceding discussion also shows why tubes are called voltage devices and transistors are called current devices. It has to do with the bias. In a tube, grid bias is accomplished with a negative voltage. In a transistor, base bias is ac-

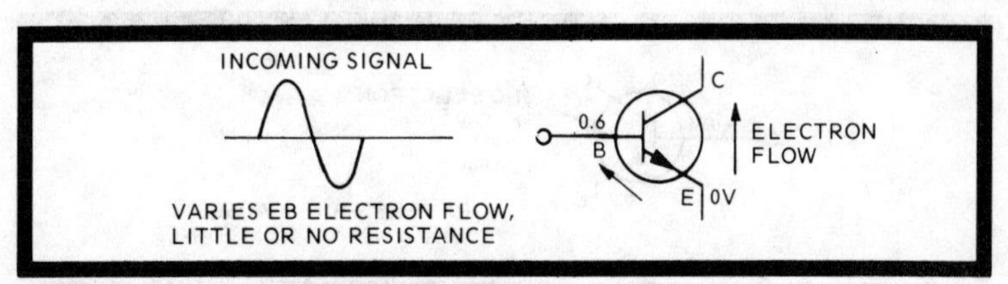

Fig. 45. The amount of EB electron flow determines the resistance between the emitter and collector.

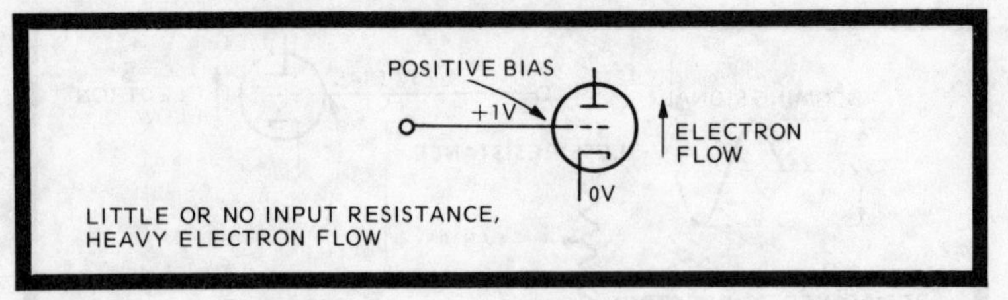

Fig. 46. When bias is positive on a tube there is little or no resistance between cathode and plate.

complished by having the EB junction turned on and current flowing. When current is flowing, a signal can be injected at the base. An increase in the peak-to-peak voltage increases the EB current flow. A decrease in peak-to-peak voltage decreases current flow. If you have enough of a decrease in peak-to-peak, the turn-on voltage can be decreased enough to turn off the EB junction and thus the transistor. Turn-on voltages are 0.6V for silicon transistors and 0.2V for germanium transistors.

Amplification

Amplification takes place in a tube as the heavy cathode-plate electron flow is modulated by a varying peak-to-peak signal voltage. In a transistor, amplification is caused by a different phenomenon. For instance, in an npn transistor, typical voltages are zero on the emitter, 0.6V on the base, and +9V on the collector (Fig. 48). Electrons come from ground, enter the piece of n material (emitter), and move toward the p material (base). Since the EB junction is forward biased, the electrons flow through easily. They encounter only a hundred or two ohms in the junction which they pass in force. A few electrons then pass into the base circuit and keep moving in the closed circuit.

The great majority of the electrons in motion, however, do not go into the base circuit. They see the large deficiency of

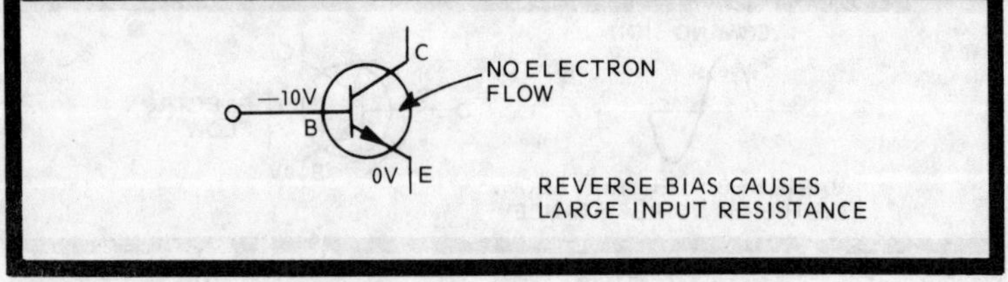

Fig. 47. A high between B and E causes a transistor to cut off.

electrons, +9V worth, at the collector and go there. In so doing they encounter the second pn junction. This junction, however, is not forward biased; there is a piece of n material for the collector with +9V on it. The p material in the base has 0.6V on it. The collector n material is a cathode and the base p material is an anode. The cathode is 8.4V more positive than the anode in this second diode of the transistor. This diode is heavily reverse biased; therefore, the junction is wide. The junction resistance is on the order of 100K.

The current passes through this junction, since the electrons are strongly attracted by the high positive voltage. As the electrons cross the high resistance, a large power gain takes place. That same modulated current that was drawn out of the emitter and passed through the EB junction flows through the BC junction. While the EB junction is a forward-biased low resistance, the BC junction is a reverse-biased high resistance. Practically all the electrons from the emitter get to the collector. Just a few are waylaid by the base circuit.

Since the power in watts is equal to I^2R, and the currents are just about the same in both junctions, the big difference in R gives a significant amount of power amplification in a transistor. Here again it can be seen that a transistor is a current-amplification device, since the current is manipulated through two pn junctions.

Most of the discussion concerned npn transistors since they are explained easier because B-plus is used on the collector like B-plus on a vacuum tube plate. Yet, pnp transistors are just as common as npn types and the servicer should be just as comfortable with them. Pnp types work exactly the same way except for the fact that electrons go to the emitter in a pnp, while electrons come from the emitter in an npn. Current flow in both the base and collector is always

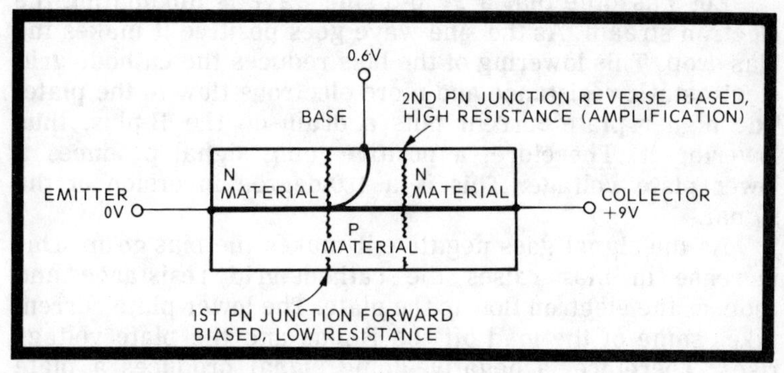

Fig. 48. Amplification in a transistor occurs because of the high resistance of the BC junction.

the same. In npn types the electrons go to the base and collector while in pnp types electrons go from the base and collector.

Thermal Runaway

In contrast to vacuum tubes, transistors cannot usually operate in excessive heat. Due to the crystalline structure of a transistor, its internal resistance decreases as it heats up. This can cause the transistor to draw more current, leading to more heat, which makes the resistance drop even more, further increasing the current demand. Once this heat buildup starts, the chain reaction can destroy the device. Watch out for heat and missing heatsinks.

25 GROUNDED-CATHODE CIRCUIT

In the tube and transistor circuits discussed so far, the cathode of the tube and the emitter of the transistor were attached to ground. This is the common-cathode or common-emitter type of circuit hookup. The word common means the element is common to both the input and output circuits. Grounded cathode and emitter circuits are used most often. As such, B-plus is applied to the tube plate. Electrons come from ground to the cathode and then on to the plate. The signal to be amplified is applied to the control grid. A battery can provide a negative voltage to the control grid. This gives the grid a bias between the cathode ground and grid.

Input Signal

Let's assume that a 2V p-p sine wave is modulating the electron stream. As the sine wave goes positive it makes the bias drop. This lowering of the bias reduces the cathode-grid electrostatic resistance and more electrons flow to the plate. The higher plate current puts a drain on the B-plus, thus lowering it. Therefore, a positive-going signal produces a lower plate voltage. This is a 180-degree inversion of the signal.

As the signal goes negative it makes the bias go up. This increase in bias raises the cathode-grid resistance and reduces the electron flow to the plate. The lower plate current takes some of the load off the B-plus and the plate voltage rises. Therefore, a negative-going signal produces a plate voltage increase. This is a 180-degree phase inversion of the signal (Fig. 49). Thus, in a grounded-cathode circuit the input signal is 180 degrees out of phase with the output signal.

Current Gain

It is evident that practically no current in the grid controls a large cathode-to-plate current flow. Since current gain is the output current divided by the input current, the current gain in a tube amplifier is great.

Voltage Gain

Here again, a small peak-to-peak voltage in the control grid circuit causes a large peak-to-peak voltage in the plate. The voltage amplification ratio is the number of times the plate peak-to-peak is greater than the grid input peak-to-peak.

These things are shown on graphs in tube manuals. While servicers can conduct repairs without understanding these graphs, knowing what they are saying provides more familiarity with the circuits and thus speeds thinking processes. A typical graph shows the operating curve of a grounded-cathode amplifier and what happens to a signal that is applied to the grid (Fig. 50). This is what you see on the scope at the tube's input and output. If you know what is supposed to be at a test point and it is missing or distorted, you have a clue.

A typical graph has the control grid bias on the base line and plate current on the vertical line. The current starts at zero and goes up to about 10 mA. The bias line starts at the cutoff bias of the tube—for instance, −10V. It goes to 0V, where the tube will usually start drawing grid current as it goes positive. Then at various bias values, a certain plate current flows. This change in bias produces a linear curve.

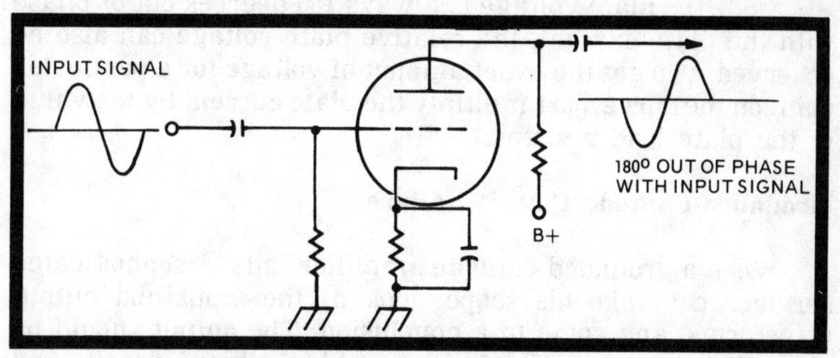

Fig. 49. A grounded-cathode circuit changes the phase of the input signal 180 degrees from grid to plate.

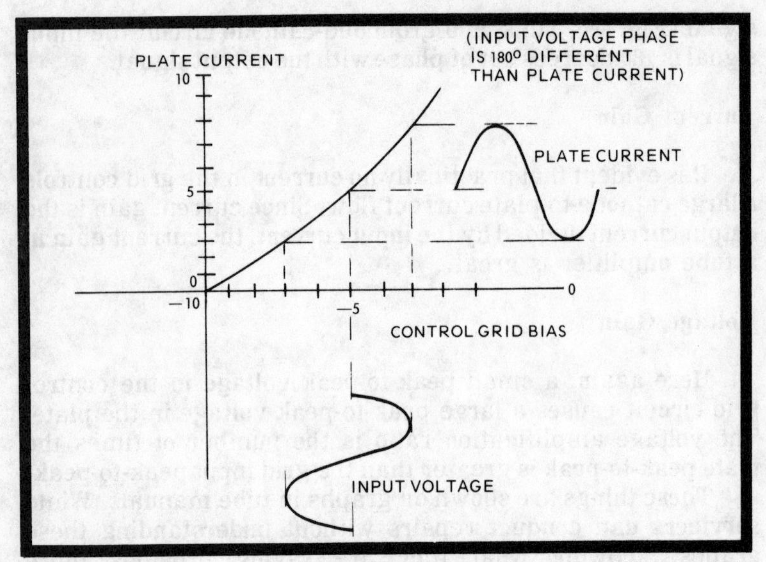

Fig. 50. Tube characteristic curve showing the effect of grid bias on plate current.

Notice that as the bias voltage goes positive, more and more plate current is produced as the cathode-grid resistance is reduced.

Next, a peak-to-peak signal voltage is applied to the bias line, which represents a signal being applied to a control grid. The peak-to-peak voltage varies the bias. For instance, a 2V sine wave varies the 6V bias between 7V and 5V. If you project the sine wave from the bias points to the curve, you can find the plate current in milliamperes for that particular signal level. Then, by simply drawing the point-by-point plate current curve produced by the signal voltage, you have projected the current amounts.

Since the plate voltage is always 180 degrees out of phase with the plate current, the relative plate voltage can also be observed. (To get the exact amount of voltage for a particular point on the curve, just multiply the plate current by the value of the plate load resistor.)

Grounded-Cathode Circuit Failure

When a grounded-cathode amplifier fails, a sophisticated servicer can take his scope, look at the input and output waveforms, and come to a conclusion. The output should be amplified a number of times, resemble the input closely, and be 180 degrees out of phase with the input. If it's not, then an intelligent appraisal of the circuit can be made. A loss or distortion of signals can indicate a particular component by deduction.

GROUNDED-EMITTER CIRCUIT 26

The grounded emitter is the most frequently used transistor circuit. In such circuits the signal is applied between the base and emitter and the output appears between the collector and emitter; thus the emitter is common to both input and output circuits.

In an npn circuit the collector receives B-plus and the base gets a positive bias. In a pnp the collector gets B-minus and the base gets a negative bias. It doesn't matter which amplifier is used, they both work exactly the same, except the electrons come from ground in an npn circuit and go to the base and collector. The electrons in the pnp start at the collector and base and go to the emitter (Fig. 51). The emitter is really not emitting in a pnp circuit but it is of no matter. The signal is still applied to the base and still comes out of the collector in both the npn and pnp circuits. The direction of electron flow has no affect on the signal. The modulation takes place either with the current or against the current. The important thing is that the EB junction is forward biased and the BC junction is reverse biased. The difference in junction resistance provides the amplification.

Input Signal

In an npn a negative-going input signal decreases the forward bias of the EB junction. As a result, the collector current decreases and the collector voltage increases. Thus, a

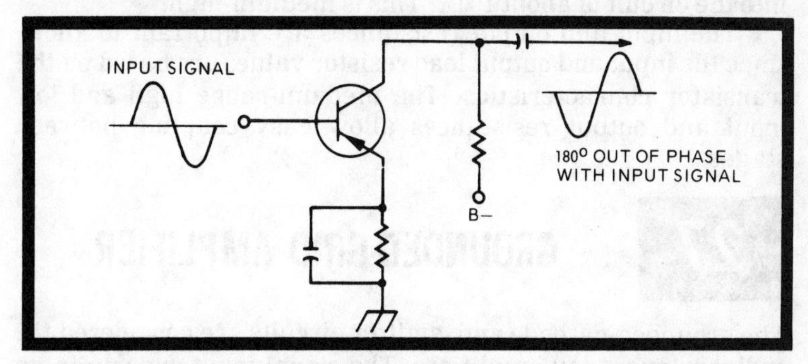

Fig. 51. A grounded-emitter circuit is similar to a grounded-cathode tube circuit. The base signal is inverted at the collector.

negative signal makes the output increase and is a 180-degree phase inversion.

Current Gain

In the transistor a small amount of current flows between the emitter and base. This small EB current affects a large current flow between the emitter and collector. Since current gain is the output current divided by the input current, the current gain is a large number.

Voltage Gain

The peak-to-peak voltage of the input signal is small and the peak-to-peak level it produces in the collector is large. Therefore, large voltage gains are made in this type of amplifier.

Power Gain

Since the power derived from the amplifier is a function of the current gain and the voltage gain, the power outputs are excellent.

Resistances

The input and output resistances are a direct function of the type of circuit hookup. That's because the resistances are a result of the way the pn junctions are arranged in the circuit. In the grounded emitter, the EB junction is forward biased and has a medium-low resistance measuring near 1000 ohms. That is its typical input resistance. The BC junction, on the other hand, is reverse biased, but not so highly biased as to spread the junction width too far. Electrons must pass through the junction. The bias causes the junction to cast a resistance into the circuit of about 100K. This is medium-high.

The input and output resistances are important to know, since the input and output load resistor values are based on the transistor characteristics. The medium-range high and low input and output resistances allow easy coupling between stages.

27 GROUNDED-GRID AMPLIFIER

The grounded-cathode and -emitter circuits are considered the ordinary everyday amplifiers. The signal input ends up in an output load with a phase change of 180 degrees. There are other ways to pass the signal with various advantages and disadvantages.

The grounded-grid amplifier (Fig. 52) is like the grounded cathode except the common point is the grid instead of the cathode. The input is applied between the cathode and grid and the output appears between the grid and plate. With the grid grounded, the signal comes in through the cathode. That way, there is no phase inversion from input to output. The signal variations are identical in phase. As the signal modulates the electron flow from the cathode a positive signal swing attracts electrons back to the cathode while a negative swing repels electrons from the cathode. In the grounded cathode, the control grid signal variation does exactly the opposite. A positive swing attracts electrons from the cathode and a negative swing pushes electrons back to the cathode.

Both types work well at low frequencies. However, as the rf range is reached, the interelectrode capacitance between the grid and plate becomes a factor. In grounded-cathode circuits large losses occur. But in a grounded-grid amplifier, the grid becomes a shield between the signal input and output points—the cathode and plate. Thus, the capacitance between the cathode and plate is very low in comparison to the grid-plate capacitances. As a result the grounded-grid amplifier is the standard amplifier in some rf applications.

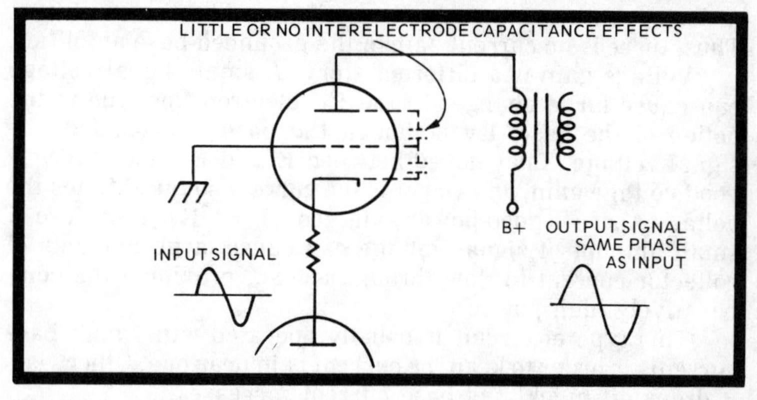

Fig. 52. In a grounded-grid amplifier, interelectrode capacitances are minimized.

GROUNDED-BASE AMPLIFIER 28

The grounded-base amplifier is the solid-state version of the grounded-grid tube amplifier. The base is the common point

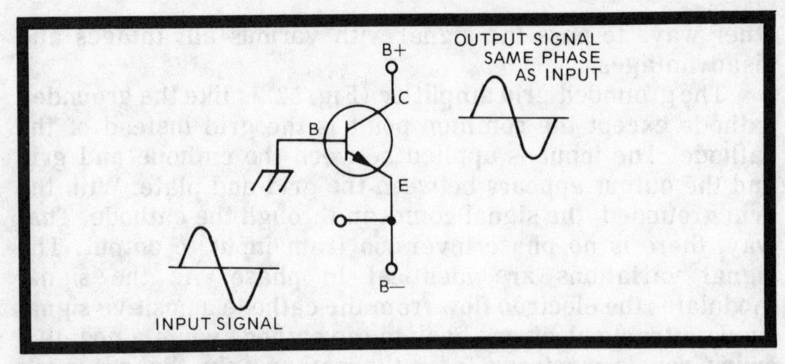

Fig. 53. A grounded-base circuit is similar to the grounded-grid circuit.

with the input applied between the emitter and base and the output available between the base and collector. Since the signal is injected at the emitter there is no phase inversion between the input and output.

In the grounded-emitter amplifier the signal is applied to the base. A small signal current change causes a large collector current change. When the signal is injected at the emitter instead of the base, the only signal current created is the emitter-to-collector current. The signal can turn the current almost all the way on. Therefore, the emitter signal current can almost cause its own value current in the collector, but not quite. Current gain doesn't quite reach one. Thus, there is no current gain in the grounded-base amplifier.

Voltage gain is a different story. A small signal voltage can cause large changes in the EC electron flow due to the action of the base. Even though the base is grounded, the signal voltage is developed between E and B. Since there is good voltage gain, and power is the current squared times the voltage, there is good power gain, too ($P = I^2R$). Therefore, a small amount of signal voltage can cause large amounts of collector current to flow through a load, providing the comparatively high power.

This type of circuit is usually operated with small base currents in order to keep the current gain near one. Otherwise, it drops off quickly as base current increases.

29 CATHODE-FOLLOWER CIRCUIT

The cathode follower could also be called a grounded plate nonamplifier (Fig. 54). With the plate grounded, the input circuit is connected between the grid and plate. The output

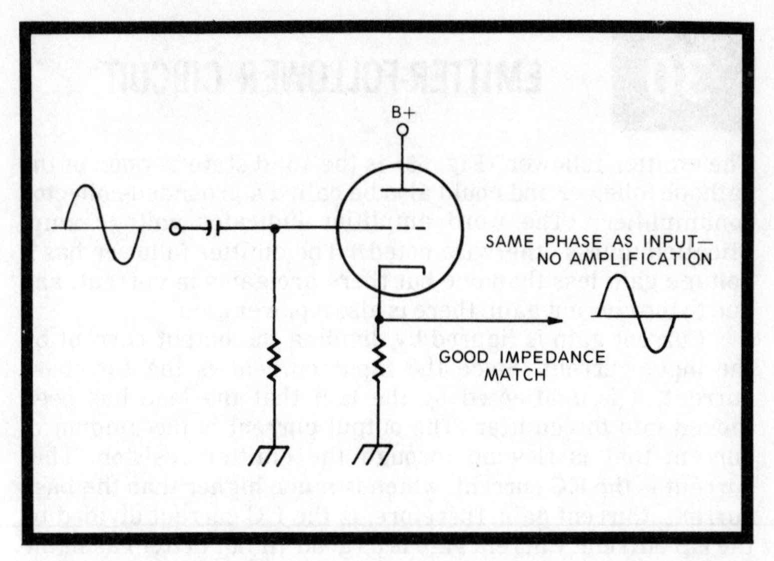

Fig. 54. Cathode-follower circuits are used to match widely different impedances.

circuit connects between the cathode and grid. There is no voltage gain; in fact, there is a loss in gain. What good is it? It's quite valuable and has many applications.

A cathode follower is actually an electron tube matching transformer. It has a high input resistance and a low output resistance. It sacrifices any gain by providing a good impedance match into a low resistance. When you have subsequent amplifiers that must handle a wide band of frequencies, the cathode follower provides an excellent wideband matching transformer.

In a cathode follower, the load is taken out of its conventional place in the plate circuit and put into the cathode circuit. The input signal is applied between the plate and grid. The dc electron flow passes from the cathode through the grid to the plate. The grid is able to change its bias in normal fashion, in accordance to the signal input, and the electron flow is modulated normally. The modulated current enters the plate, but there is no load in the plate. The plate is grounded (through the power supply). The electrons go through ground, then back into the cathode. There the electrons encounter the load resistor. The voltage drop across the load follows the input signal; that is, there is no phase inversion. The cathode resistor can be adjusted in value to match the input of the subsequent amplifier. All of the frequencies are passed with little or no distortion.

30 EMITTER-FOLLOWER CIRCUIT

The emitter follower (Fig. 55) is the solid-state version of the cathode follower and could also be called a grounded-collector nonamplifier. (The word amplifier indicates voltage amplification unless otherwise noted.) The emitter follower has a voltage gain less than one but there are gains in current, and due to the current gain, there is also a power gain.

Current gain is figured by dividing the output current by the input current. Since the input current is the tiny base current, it is unaffected by the fact that the load has been moved into the emitter. The output current is the amount of current that is flowing through the emitter resistor. That current is the EC current, which is much higher than the base current. Current gain, therefore, is the EC current divided by the EB current. Current gain is as good (if not better) as in the grounded-emitter amplifier. And with the good current gain, the power gain is also high.

The principal application of the emitter follower is as an impedance-matching transformer. For instance, a high-impedance microphone can be attached to the input of an emitter follower. Then the output of the follower is injected into a conventional amplifier with a grounded-emitter input. The emitter-follower resistor can then be a few hundred ohms and match the 100K output of the microphone into the low input resistance of the amplifier. There are many common uses you'll encounter during servicing.

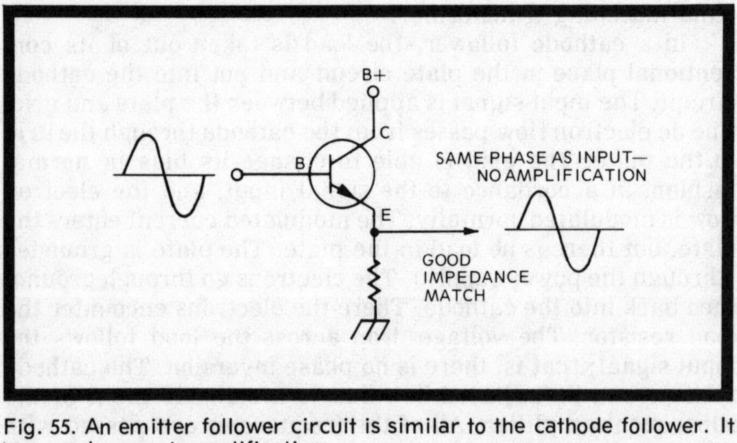

Fig. 55. An emitter follower circuit is similar to the cathode follower. It has good current amplification.

RESISTANCE-COUPLED TUBE CIRCUIT 31

In both vacuum tube and transistor amplifier circuits, there must be a **load**. This is the end of the circuit, where the actual work is performed. It can be a loudspeaker, crt, switch, etc. It also can be a resistor (Fig. 56). Each amplifier has its own load. When it is a resistor, the amplified signal appears across it in the form of a voltage drop (Fig. 57).

There are two separate voltages in an amplifier circuit. One is the dc from the power supply. It is simply a continuous flow of electrons through the circuit. The second is the ac component on the signal. The ac component causes the electrons to vary in intensity.

The ac component is developed across the plate load resistor. In a tube circuit the B-plus voltage across the resistor could be 120V on the power supply side and then drop to 100V on the plate side. There is a 20V drop. This is a drop between the plate and the cathode in the grounded-cathode circuit. The ac voltage causes the electron flow to vary through the resistor. These variations are much larger than the ac voltage input between the grid and cathode.

The amplified variations have to be passed into the next stage. However, the input to the next stage may have —5V on the grid. If the 120V B-plus is applied, it will swamp out the next stage bias. So a coupling capacitor is placed in series. It blocks the B-plus but easily passes all of the ac variations (Fig. 58).

The ac output of the preceding stage is then developed across the grid input resistor. The resistor has to be much

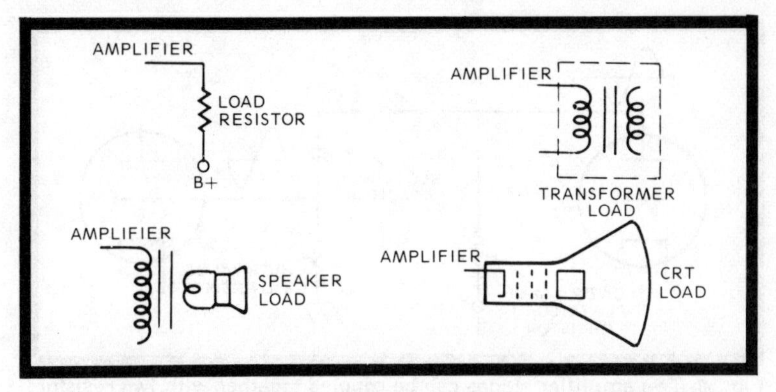

Fig. 56. An amplifier delivers its output to a load of some type.

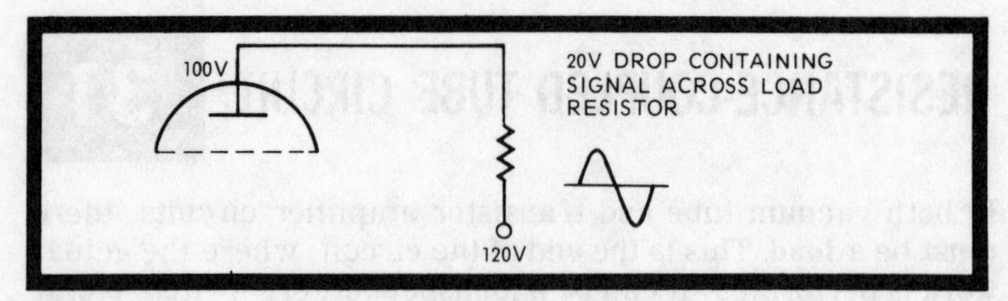

Fig. 57. A load resistor has a voltage drop across it. The drop develops a portion of the signal.

larger than the preceding plate resistor since the two resistors are in parallel as far as ac is concerned, even though they are blocked off from each other as far as dc is concerned. If the grid resistor were lower in value, it would sap off some of the ac amplification appearing in the plate resistor. The two resistors are the input resistance to the second amplifier as well as the output resistance to the first amplifier. This type of coupling is common and must be understood completely for expert servicing.

Resistance-Coupled Circuit Tests

The plate and grid of the typical grounded-cathode amplifier are the key test points during circuit failure. If you want to trace the signal, you use a scope, set it at the signal frequency, attach a high-impedance probe, and touch it to the grid and plate of each stage. The signal appears on the scope and you can follow it from stage to stage. As soon as you lose it, you have passed over the defect area.

If you want to inject a test signal, the grids and plates are the best injection spots. Starting from the last stage and

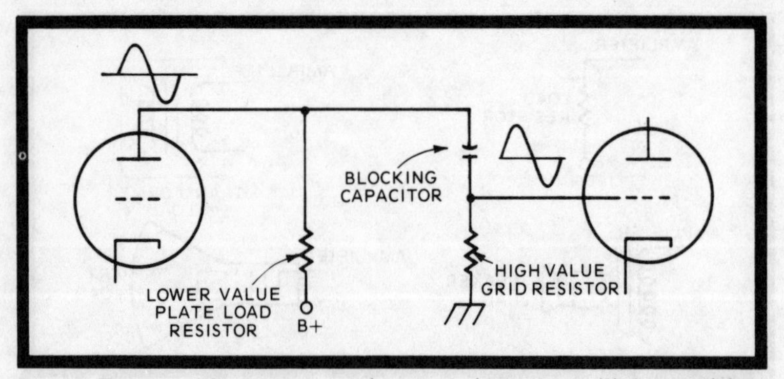

Fig. 58. Two amplifier stages can be coupled together with two resistors and a capacitor.

working forward, the test signal is injected until it disappears. When that happens you have passed over the defect area.

Once the defective stage is isolated, that plate and grid become dc test points. Ac signal tracing or injection is used to isolate the defective stage. The dc test points help to pinpoint the actual defective component.

No B-Plus

The first step is to test the plate for B-plus (Fig. 59). With the gear turned on, a voltmeter probe is touched to the plate. Now you must analyze the B-plus voltage. Let us suppose that the power supply is providing 120V and under normal operation 20V is dropped across the plate resistor. The schematic shows that 100V should be present at the plate.

One defect could show 0V on the plate. What could cause that? First of all, the plate load resistor could be open. If that happened, the power supply could not cause a deficiency of electrons on the plate and pull electrons from the plate through the load resistor.

The resistor is tested with an ohmmeter. If it's bad, of course, it must be replaced. Suppose it's good? What could have killed B-plus on the plate? There is a plate bypass capacitor to ground. It could have shorted. If it did, then the power supply is pulling electrons across the load resistor, but from ground directly through the shorted capacitor. Also, since the current drain is so strong through the resistor, the voltage drop reaches 120V (E equals IR). The current drain

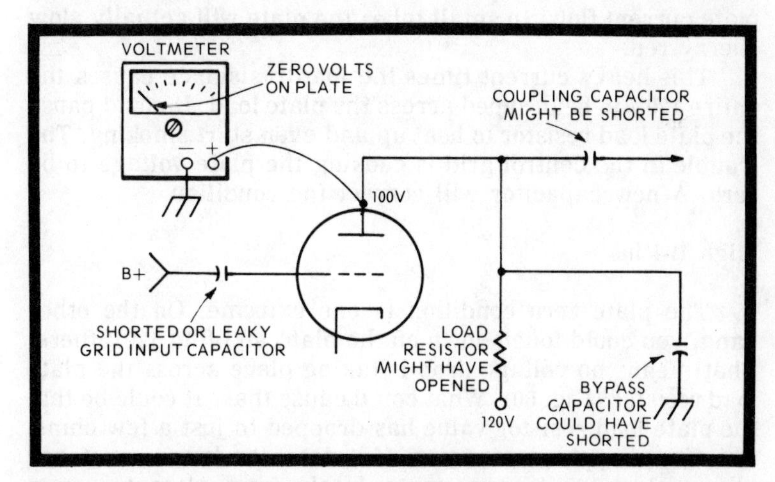

Fig. 59. When a plate reads zero instead of B-plus, one of four common component breakdowns may be responsible.

has become large. Multiply it times the load resistance and 120V is the result. With a 120V drop across the resistor, no voltage is left for the plate.

But suppose the capacitor is good, too? What else could cause the plate to read 0V dc? There is the coupling capacitor to the next stage. If it shorts and there is a very low voltage to ground in the next stage input, the plate would read 0V. The shorted coupling provides a low-resistance path to ground. The B-plus could attract electrons from ground through the coupler and back across the plate load resistor to the power supply. Again the voltage drop across the plate load would be 120V and rob the plate of B-plus.

Suppose the coupler is good? There are no other components in the circuit. It's unusual for the wiring to develop a short to ground or an open. It could happen, especially on printed boards, but it is unusual. A visual inspection shows all the wiring to be good.

The next suspect must be the tube. However, it was replaced right off and a known good tube is in the circuit. Next area for inspection is the grid circuit.

Suppose the grid input capacitor has shorted. That would put the plate B-plus from the previous stage on the control grid. B-plus on the control grid turns the tube on full blast. It overrides the negative bias voltage completely and forms a powerful attraction to the cathode space cloud. Some of the electrons go directly to the grid and are observed as a heavy damaging grid current. As fast as the electrons are emitted from the cathode, they fly to the plate, resulting in a heavy plate current flow. In small tubes the plate will actually glow cherry red.

This heavy current times the load resistance causes the entire 120V to be dropped across the plate load. It could cause the plate load resistor to heat up and even start smoking. The trouble in the control grid is causing the plate voltage to be zero. A new capacitor will correct the condition.

High B-Plus

The plate zero condition is one extreme. On the other hand, you could touch down on the plate and find 120V there. That means no voltage drop is taking place across the plate load resistor (Fig. 60). What could cause that? It could be that the plate load resistor value has dropped to just a few ohms. This does happen sometimes. If it does, the lower resistance allows electrons to pass more freely from plate to power supply. The less the resistance the less voltage drop there is. If the resistance is near zero, then the electrons pass freely through the resistor and there is no voltage drop. So the plate reads 120V. The resistor can be tested with an ohmmeter and if it is defective, it can be replaced.

The rest of the plate components in this case can be more or less exonerated. If one of them or the wiring somehow causes a short across the plate load, the subsequent short circuit could provide a low-resistance path for the electron flow. There would be no voltage drop as a result. This possibility is not too likely.

If the plate resistor is good, then the reason for the loss of voltage drop is due to the tube not conducting. Without normal electron flow in the tube, no electrons are passing through the load resistor and no voltage drop.

The tube could be cut off by troubles in the cathode or control grid. It could be a higher than normal grid bias or an open cathode circuit. A higher grid bias can happen if there is a B-minus voltage nearby that accidentally gets into the grid due to a short circuit.

Also, a higher grid bias can occur if too much signal is applied to the blocking capacitor, in which case the high signal level charges the capacitor to a high level. The result is a high negative charge. As the signal varies, the capacitor charge tends to leak off to ground through the grid resistor at a rate

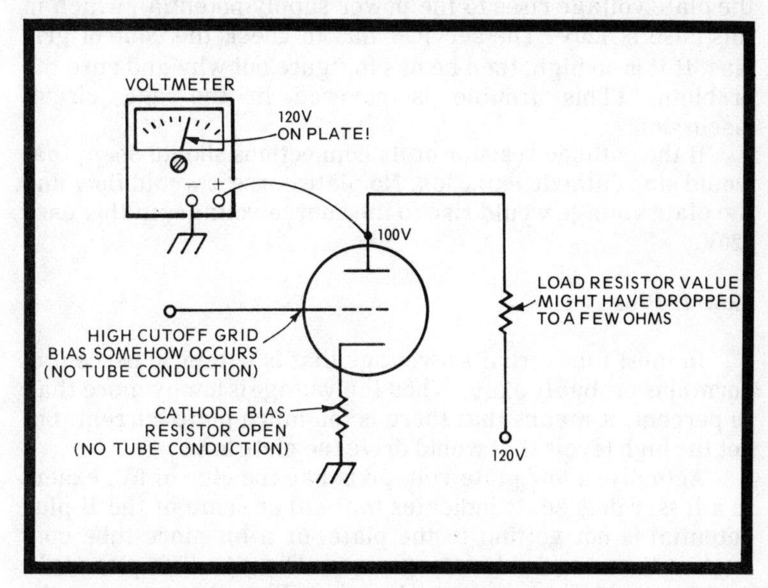

Fig. 60. When a plate reads full B-plus instead of its prescribed voltage, it may be due to one of three causes.

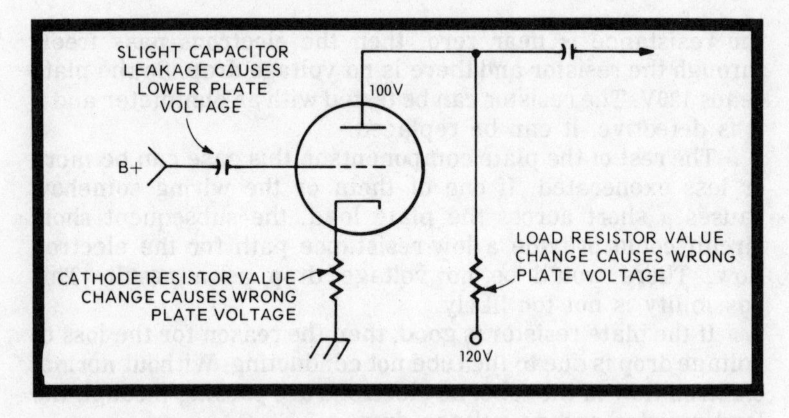

Fig. 61. When a plate voltage reads incorrectly, it could be due to one of the three reasons shown.

determined by the time constant of capacitor and resistor. If the signal is much higher than normal, not too many electrons get a chance to trickle off to ground, leaving a negative charge on the grid at all times. This negative charge, if higher than the tube's cutoff value, stops electron flow in the tube. When the tube is cut off and no electrons flow from cathode to plate, the plate voltage rises to the power supply potential, which in this case is 120V. The servicer has to check the control grid bias. If it is so high, then he has to figure out why and cure the problem. (This trouble is covered in the agc circuit discussion.)

If the cathode resistor or its connections should open, that would stop cathode emission. No plate current would flow and the plate voltage would rise to the source voltage, in this case 120V.

Low B-Plus

In most tube circuits a voltage that is within 20 percent of normal is probably okay. When the voltage is low by more than 20 percent, it means that there is too much plate current, but not the high levels that would drive the plate to 0V.

Actually, a low plate voltage is like the clue of 0V, except to a lesser degree. It indicates that either some of the B-plus potential is not getting to the plate, or a bit more tube conduction than normal is taking place (Fig. 61). The plate load resistor could have increased in value. When that happens, the increased resistance doesn't let as many electrons flow from the plate to the power supply. As a result the B-plus on the plate is reduced. In other cases, if the plate resistor value decreases, the B-plus can increase on the plate.

Now to confuse you altogether, you will also encounter cases where the B-plus is low and you find the plate load resistor value has dropped. How can that be? The lower plate resistance allows more electrons to leave the plate than normal. This should produce a larger deficiency, tending to make larger values of B-plus. However, in some instances the power supply is marginal in ability to hold its voltage. As the additional electrons pour into the power supply, the source positive voltage goes down as the electrons add their negative charges to the power supply. Actually, the plate voltage itself is not being affected as much as the source voltage. The B-plus in the entire circuit decreases.

The way to test for this complication is to check the voltage on both sides of the plate resistor. If B-plus is normal on the source side, then the power supply is okay and the lower plate voltage is due to the components. Should the source voltage be low, then look into the reason for that trouble. There are some power supply discussions further on in the book.

Other cases of in-between B-plus readings could be due to the control grid or cathode components. If the grid blocking capacitor should leak a little and let a volt or two into the grid circuit, the positive voltage will lower the negative bias, causing more conduction, which lowers the plate voltage. If the cathode resistor decreases in value, the grid-cathode bias relationship is changed in a positive direction, also causing more conduction and a lower plate voltage.

The important thing in servicing the tube circuit is "seeing" the electrons as they flow through the circuit. They come from ground and enter the cathode circuit. There they are slowed somewhat by the cathode resistor, then are emitted into the space cloud. The cloud looses some electrons as the nearby control grid dictates by its changing negative charge and the plate attracts the electrons. They journey to the plate, enter the plate circuit and encounter the plate blocking capacitor and plate load resistor.

At the plate blocking capacitor, the electrons induce electrons on the other side of the dielectric to move in step with their ac movement. Then they pass through the plate load resistor and go on to the power supply and back to ground.

The electrons that were caused to vibrate in time on the other side of the capacitor dielectric move into the next stage control grid, causing the grid bias to vary according to the ac signal.

32 RESISTANCE-COUPLED TRANSISTOR CIRCUIT

A transistor circuit amplifier has a load resistor, too. It is called the collector load. When the load is a resistor, it acts like a plate load resistor, but there differences are due to the transistor characteristics.

A collector has a typical operating voltage around 9V. This is quite small in comparison to the tube's 120V. Also, most transistors are tiny in comparison to a tube and the collector current is small in comparison to plate current. This is a definite advantage and permits miniaturization and very little heat buildup. Since heat expended in electronic gear is all wasted energy, great advantages are gained.

These tiny voltages, currents, and physical sizes, while great for equipment, make it harder for the servicer. Instead of having considerable space to gain access to test points and free soldering room, transistors are located in small spaces. Tube circuits are like an alarm clock, while the transistors are analogous to a tiny wrist watch.

The collector voltage can be compared to the plate voltage. In an npn the polarities are the same; the collector voltage is positive like plate voltage. In pnp transistors, however, the polarities are opposite; the collector voltage is negative. Instead of traveling from emitter to collector, electrons travel from collector to emitter. But regardless of dc electron flow, the signal can change it.

The base is polarized in the same way as the collector. Npn types have positive bases while pnp types have negative bases. That is, of course, if the EB junction is forward biased, which is considered the conventional mode. If the transistor is reverse biased, the base polarity is opposite to the collector.

Transistor Cut Off Test

For transistors there is a convenient test that cannot be used with tubes. It is easy for the servicer to cut off the transistor. All you have to do is short the emitter to the base (Fig. 62). The short places the same voltage on both emitter and base. Since a silicon transistor must have 0.6V forward bias to turn on (0.2V if it's germanium), by making the two points the same voltage the forward bias is eliminated and the transistor cuts off. This, as you'll see, is a valuable testing technique. (The only problem is if you should accidentally get some form of voltage charge on the base or emitter during the test and rupture the EB junction. A new transistor will be needed.)

In a tube you could accomplish cutoff by adding a high negative bias to a control grid. This would increase the grid bias enough to cut off the tube. Unfortunately, even though some good tests could be realized with the technique, it is too much trouble. There are other convenient tests. In a transistor, though, the cutoff technique is one of the mainstay tests a servicer makes.

Let's see what the cutoff tests can prove. Typically, a transistor power supply puts out a potential of 12V. By the time the voltage appears on a normally conducting transistor, the collector reads 9V. Suppose you suspect a transistor as defective. All you have to do to perform a quick in-circuit check is take a reading at the collector. If it reads 9V, you then short E to B and read the collector again. The short turns off the transistor. With no EC current flowing, the collector voltage should rise to the 12V source if the transistor is good. If it doesn't, the transistor is leaking. There is current flow in the transistor even though there is no forward bias. The only place the current could go is through ruptured junctions.

When the source voltage does appear on the collector as the EB junction is shorted, chances are good the transistor is perfect. At any rate you know the valve action is fine. When the transistor turns on and off normally, it is usually good. The nice part of the short test is the ease with which it can be accomplished.

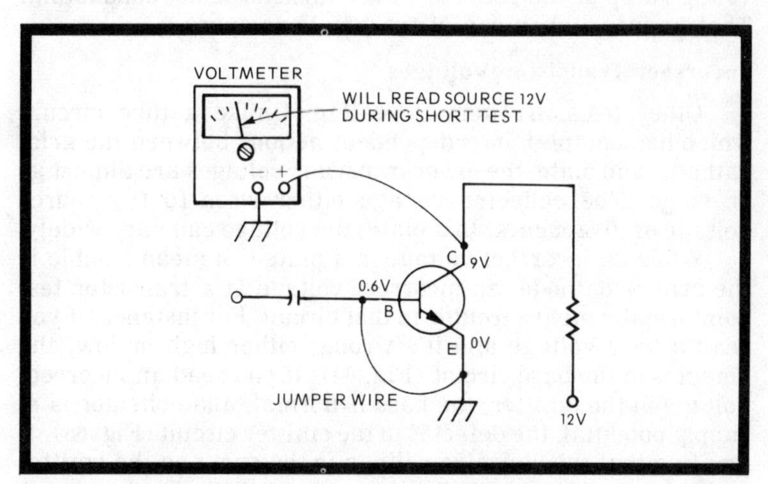

Fig. 62. A jumper wire is used frequently in transistor service work since it can be used to cut off the device.

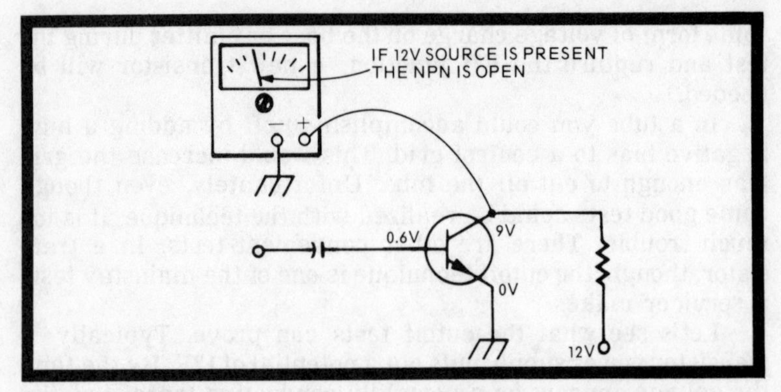

Fig. 63. When npn bias is correct, but the collector reads full B-plus, the transistor is open.

Open Transistor Test

A transistor is easily checked for an open junction also. If the transistor is forward biased, the collector is a few volts below source as collector current flows. If you suspect that a transistor is open, all you have to do is take careful voltage readings of the collector, base, and emitter. If the emitter and base have the correct forward bias (0.6V for silicon and 0.2V for germanium) but the collector is at the source voltage, the transistor is open (Fig. 63). The correct forward bias turns on a good transistor and causes a voltage drop across the collector load resistor. When the bias is correct and there is no voltage drop at the collector, the transistor is not conducting. That happens when one of the junctions opens.

Incorrect Transistor Voltages

Other tests are very revealing. Unlike a tube circuit, which has complex interdependent actions between the grid, cathode, and plate, the lower transistor voltages are almost go or no-go. The collector voltage either rises to the source voltage or disappears. In a plate, the voltage can vary widely.

While an incorrect voltage in a plate can mean trouble in the grid or cathode, an incorrect voltage at a transistor test point usually means trouble in that circuit. For instance, if you read a base voltage and it's wrong, either high or low, the defect is in the base circuit (Fig. 64). If you read an incorrect voltage in the emitter, the base is normal, and collector is at supply potential, the defect is in the emitter circuit (Fig. 65). If you find that the collector voltage is the same as the emitter and the base is normal, the defect is in the collector circuit (Fig. 66).

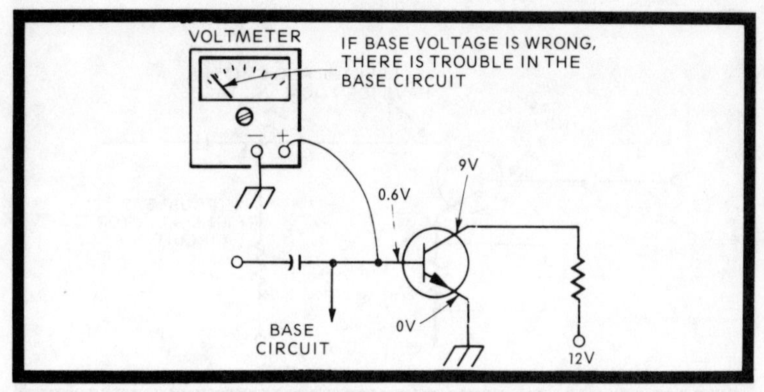

Fig. 64. When the base voltage is wrong, there is trouble in the input circuit.

When a particular circuit leg is indicated, the components must be tested one by one with the equipment off. During the test, be sure damaging voltages are not applied to the transistors. Many ohmmeters have batteries large enough to rupture a junction. If you are using a filter capacitor for any tests, be sure you discharge it before installing it in a circuit. Otherwise, the charge could accidentally enter the wrong side of a pn junction and blow it.

IMPEDANCE-COUPLED CIRCUIT 33

Resistance-coupled amplifiers are valuable in that they give a satisfactory response over a wide range of frequencies. However, there is a major disadvantage in that the voltage

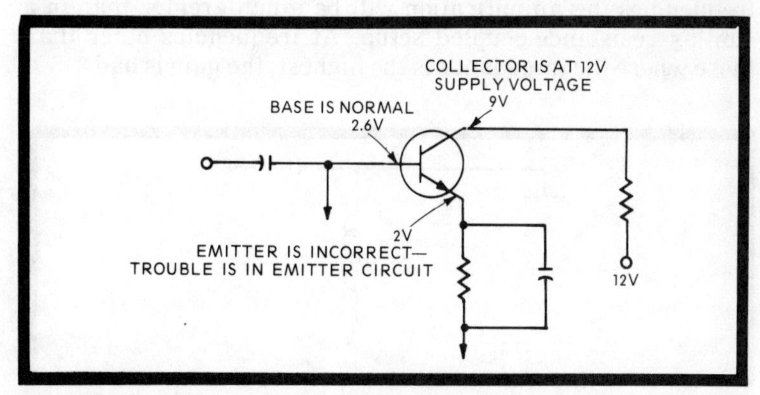

Fig. 65. When the base is normal, collector is at full B-plus, and emitter voltage is off, there is trouble in the emitter circuit.

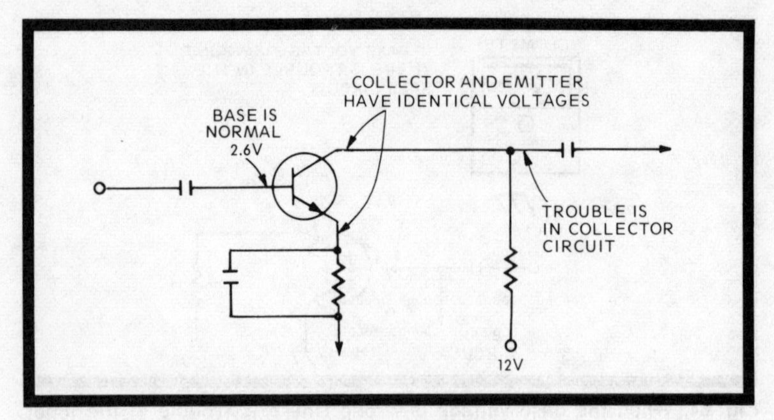

Fig. 66. When the base is normal but collector and emitter voltages are identical, it is an indication of collector circuit trouble.

drop across the resistance can be wasteful. The power supply voltage must be high enough to afford the voltage drop and still put enough B-plus on the plate or collector to power it.

In order to reduce the necessary supply voltage, impedance coupling can be used (Fig. 67). It is identical to resistance coupling except a coil is used instead of the plate resistor. The coupling capacitor and grid resistor are still present. The coil acts like a piece of wire to the dc of the power supply and practically no voltage drop takes place. But the coil is an inductance to the ac being amplified. And since it has a definite impedance at the ac signal frequency, the coil is a load.

According to the number of turns in the coil, the load it presents is sensitive to certain frequencies. At these frequencies the amplification will be much greater than in a similar resistance-coupled setup. At frequencies other than those where the impedance is the highest, the gain is bad.

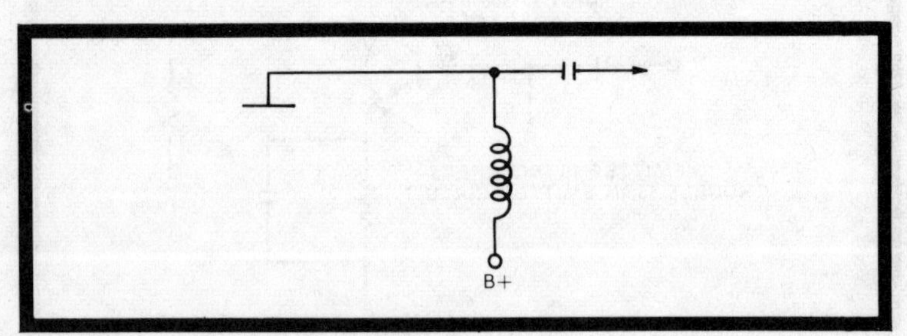

Fig. 67. Instead of a plate load resistor, a coil can be used. It must have a high impedance at the operating frequency.

Impedance-Coupling Circuit Failure

The same testing techniques are used on this circuit. The coil itself tends to open. On occasion, when windings are on top of each other, they can short together through the insulation. When that happens the best test is to try a replacement coil. Resistance readings show little.

TRANSFORMER-COUPLED CIRCUIT 34

A coupling transformer does the job of the plate load, grid input resistor, and blocking capacitor. The plate load is the primary, the grid input is the secondary, and the dc blocking action is provided by the separate windings (Fig. 68). The dc can't pass from primary to secondary but ac does easily. The number of turns in the primary and secondary determines the impedance. By selecting the correct inductances, any impedance can be installed in the primary and the secondary. The two impedances work together perfectly.

It is a fact that the least loss takes place when the signal source matches the load. If a transistor with an output resistance of 50K is feeding a second transistor with an input resistance of 2K, a transformer can be used to match the output and input impedances. The primary of the transformer has an impedance of 50K at the signal frequency and the secondary of the transformer has an impedance of 20K. The primary will have 5 or 10 times the number of turns in the secondary.

In transistor coupling, the transformers are always stepdown types, since the input of a transistor is such a low resistance. Tubes, on the other hand, have a high input resistance and you'll find coupling transformers that are either stepup or stepdown, according to the resistance of the load the transformer has to feed.

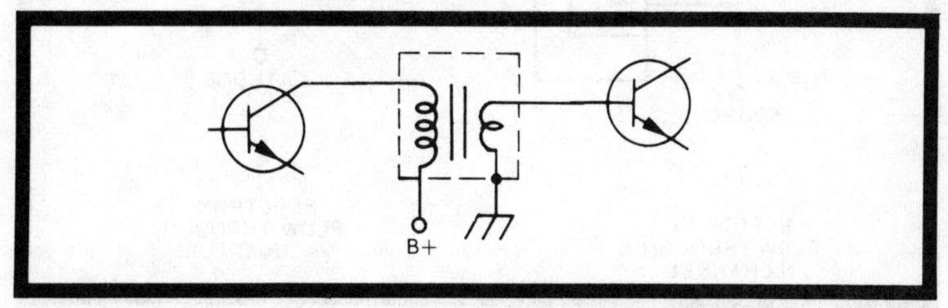

Fig. 68. A transformer can be used to couple a high-impedance output to a low-impedance input.

35 FET COUPLING CIRCUITS

When transistors are mentioned, the servicer has to remember to think of the input across the base and emitter as a low resistance. This is true for all npn and pnp transistors. It is not true with FETs. Field-effect transistors, even though they use pieces of p and n material as basic building blocks, do not act like ordinary transistors. There is no pn junction for the electrons to cross. There is a pn junction, but it does not perform the same as ordinary transistor junctions.

In an n-channel FET the current enters one end of a piece of n material and leaves the other end (Fig. 69). During its journey through the n material, the electrons pass a pn junction and are restricted or let alone, according to the electrostatic charge on the other side of the junction. No electrons cross the junction. It is always biased to avoid that.

Yes, that is exactly the way electrons pass through a vacuum tube, entering the cathode, leaving via the plate after being affected by an electrostatic charge on the control grid. The FET is quite the same as a vacuum tube in characteristics. Only thing is the electrons are passing through a

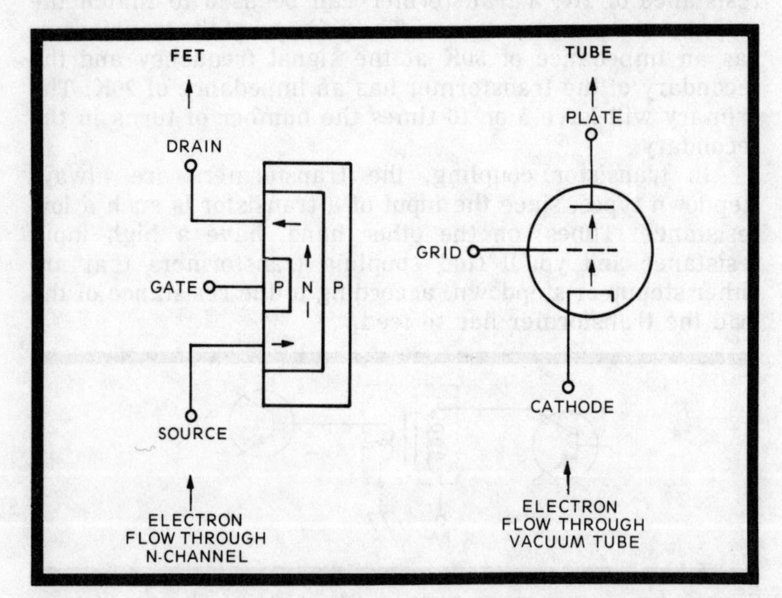

Fig. 69. An FET has many triode tube characteristics without the need for a heater.

piece of n or p material (n-channel or p-channel) instead of through a vacuum.

As a result, the input to an FET is like the input to a tube—a high resistance. Any coupling that is used is quite similar to tube coupling. The FET combines the advantages of the tube and transistor and has high input and output resistances.

CLASS A TUBE AUDIO AMPLIFIER 36

The typical Class A audio amplifier is found in a radio or TV after the audio detector and before the audio output. An audio amplifier is a voltage amplifier. It amplifies the weak detector output into a larger voltage (Fig. 70).

Class A is the mode of operation where all of the input signal appears in the output with no clipping of peaks. The Class A operation is attained by setting the grid bias in a tube or the base bias in a transistor so that the input signal operates over the straight portion of the characteristic curve of the device (Fig. 71).

The detector output is developed across a load resistor, usually a variable type called the volume control. In a tube circuit the control is a voltage divider. The voltage divider is fed signal from a low-resistance source (the detector) and operates into a high-resistance (the grid input). The audio voltage is developed across the portion of the control that is not bypassed. When the control is turned all the way up, all the

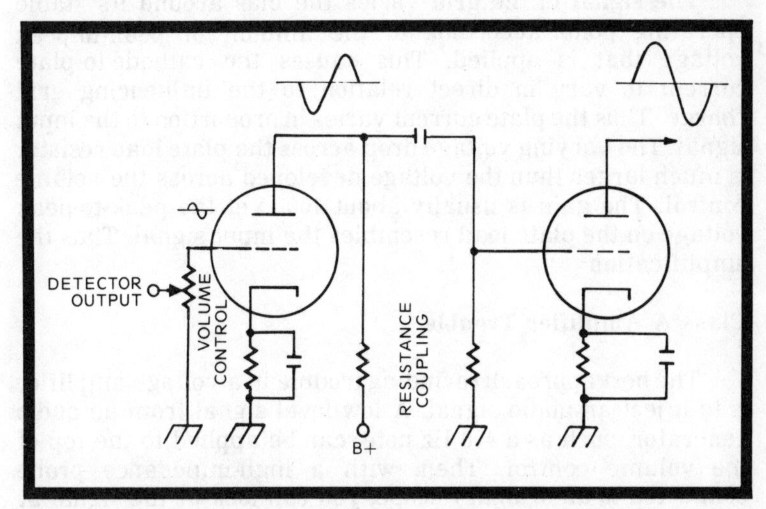

Fig. 70. A Class A amplifier processes a signal from one stage to another with little or no distortion.

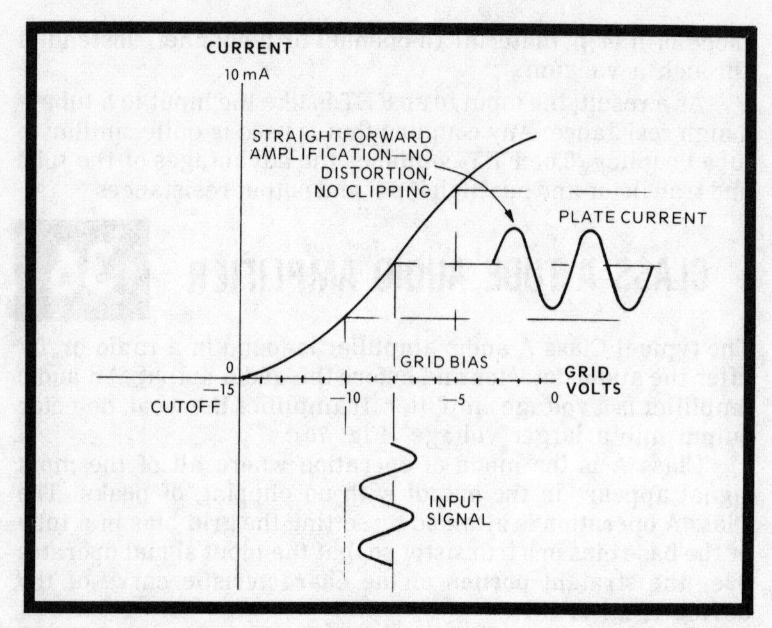

Fig. 71. A Class A amplifier is biased to operate on the linear part of the characteristic curve.

available audio from the detector is used and the amplifier receives the maximum signal input. When the control is turned down, smaller and smaller amounts of signal get to the amplifier.

The signal in the grid varies the bias around its stable operating point according to the amount of peak-to-peak voltage that is applied. This causes the cathode-to-plate current to vary in direct relation to the influencing grid charge. Thus the plate current varies in proportion to the input signal. The varying voltage drop across the plate load resistor is much larger than the voltage developed across the volume control. The gain is usually about 100. Yet the peak-to-peak voltage on the plate load resembles the input signal. Thus the amplification.

Class A Amplifier Trouble

The best approach to finding trouble in a voltage amplifier is to inject an audio signal. A low-level signal from an audio generator, such as a 400 Hz note, can be applied to the top of the volume control. Then, with a high-impedance probe connected to an ordinary scope, you can look at the signal at the plate end of the load resistor. If the sine wave cannot pass through the circuit, the stage is obviously dead. Should the sine wave be present, the stage has some gain. Gradually turn up the output gain of the generator and the volume control and watch the scope waveform. The sine wave should, in a good circuit, simply increase in amplitude up to a point where its amplitude exceeds the characteristic capabilities of the tunbe. At that point the top or bottom or both of the waveform peaks will start to flatten out, as the signal either reaches the tube cutoff or transistor saturation.

If the sine wave is distorted or has other queer appearances, the grid bias isn't set properly or the plate load is not correct. Check the few resistors, capacitors, and other components.

CLASS A TRANSISTOR AUDIO AMPLIFIER

37

The typical Class A transistor audio amplifier is used for the same purposes as its tube-type counterpart. Its input is also from a volume control but the control is a current divider. The current divider is driven from a high resistance, such as the collector of a transistor, and feeds into a low resistance on the EB junction of the transistor amplifier (Fig. 72). Since the bias on the base is a result of a current flow between emitter and base, varying the bias actually amounts to varying the amount of EB current.

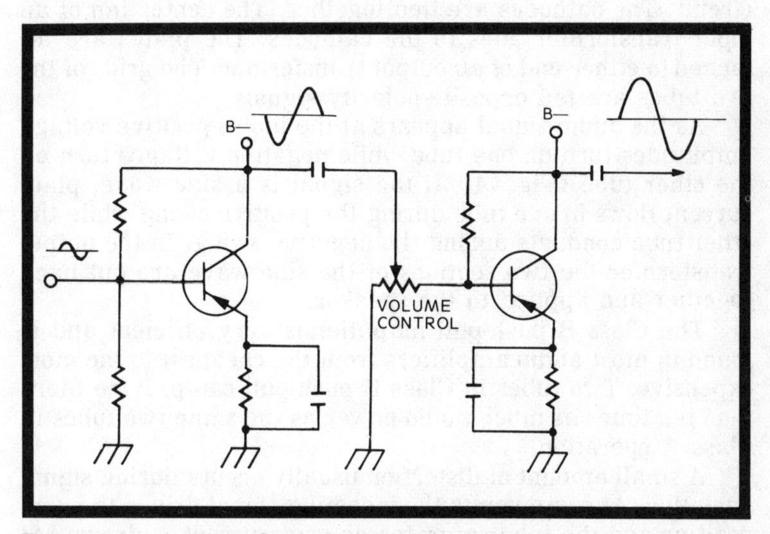

Fig. 72. A transistor Class A amplifier, for all practical purposes, processes a signal exactly like its tube counterpart.

Class A Amplifier Trouble

Class A operation in a transistor, although determined by an EB current instead of a control grid voltage, is still the same. All of the input signal is developed across the collector load resistor with an amplification of about 100. A signal generator and scope can be used to troubleshoot this circuit, following the procedure suggested for the tube circuit. The only differences are the lower operating voltages and the few circuit changes. The emitter-to-collector current flow is like the cathode-to-plate current flow and produces the same kind of voltage amplification in the collector load resistor.

38 CLASS B PUSH-PULL AUDIO AMPLIFIER

Class B amplifiers are biased at cutoff instead of in the middle of the conduction range like a Class A amplifier. A Class B amplifier is cut off when no signal is applied. This is a valuable characteristic in battery-operated circuits since no plate current is drawn when no signal is applied to the input. (Class A amplifiers are always conducting because the bias is set in the center of the conduction range.)

Class B amplifiers are usually employed in pairs in a push-pull circuit configuration (Fig. 73). The plates and grids of the two tubes are connected to opposite ends of a balanced circuit. The cathodes are tied together. The center tap of an input transformer goes to the cathodes. The plates are attached to either end of an output transformer. The grids of the two tubes are fed opposite-polarity signals.

As the audio signal appears at the input, positive voltage amplitudes turn on one tube while negative voltages turn on the other tube (Fig. 74). If the signal is a sine wave, plate current flows in one tube during the positive swing while the other tube conducts during the negative swing. In the output transformer the two sections of the sine wave are put back together and applied to the speaker.

The Class B push-pull amplifier is very efficient and is found in most audio amplifiers from the cheapest to the most expensive. Two tubes in Class B push-pull can provide more than ten times as much audio power as the same two tubes in Class A operation.

A small amount of distortion usually occurs during signal variation. At some points the incoming signal drives the grid positive and the tube saturates as grid current is drawn. At lower levels of volume the distortion is kept at a minimum.

Class B Push-Pull Trouble

When large volume levels are passed through a Class B push-pull amplifier, the input resistance makes drastic changes. During the part of the signal that is low in amplitude, no grid current is drawn and the effective load resistance is high. Little strain is placed on the circuit. At the next instant a large signal turns the grid into the positive range and current is drawn. The load resistance drops to a very low value. The strain is on the tubes, input and output resistors, and the previous audio amplifier (driver) stage. They can fail.

The circuit can be repaired easily with an audio signal injected at the input. To isolate the defective stage, pull one of the tubes and listen to the speaker. If you pull a tube (in a series circuit you'll have to do it quickly before the heaters go out), the entire stage goes dead. Therefore, the remaining tube circuit is dead.

Once the defective stage is isolated, dc voltage and resistance readings should be checked. Test for the presence of B-plus voltage and check individual components when incorrect voltages appear.

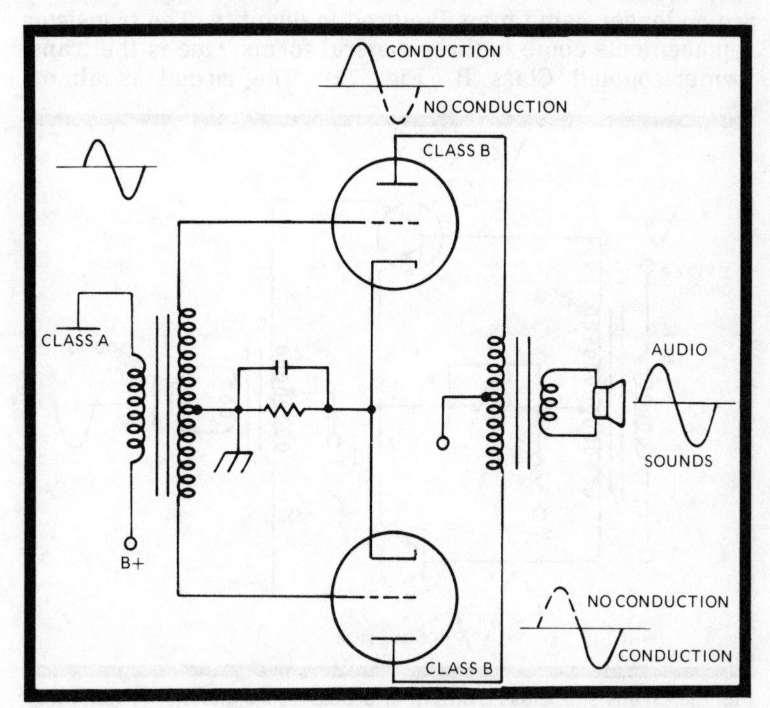

Fig. 73. A Class B amplifier is typically used in a push-pull configuration.

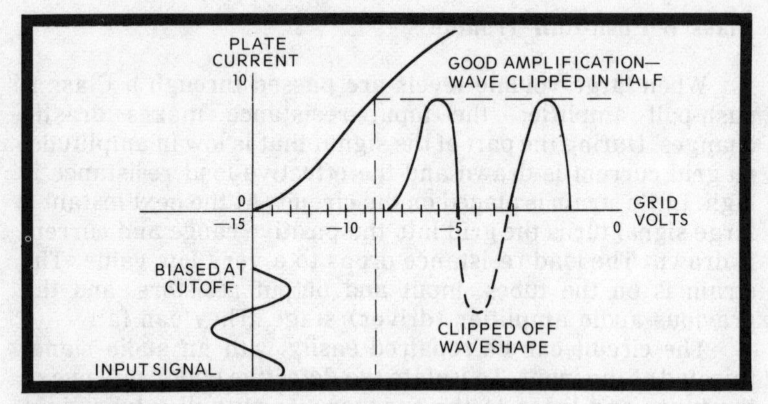

Fig. 74. A Class B amplifier stage is biased at cutoff. Only the signal above cutoff is amplified.

39 CLASS B TRANSFORMER-COUPLED (PUSH-PULL) TRANSISTOR

While tube Class B push-pull type audio amplifiers are still around and will have to be serviced for a time to come, they are no longer being manufactured in quantity. The transistor replacements come in three general forms. One is the transformer-coupled Class B (Fig. 75). The circuit is almost

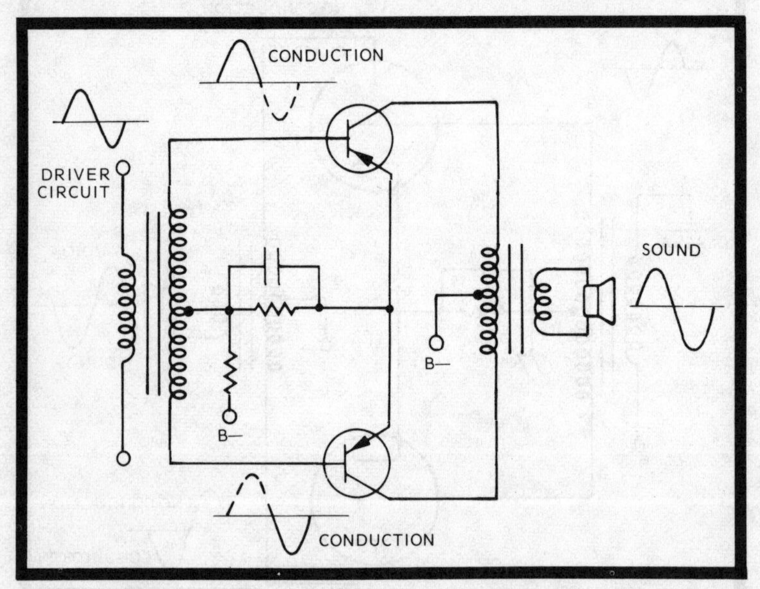

Fig. 75. A transistor Class B amplifier processes a signal in the same way as a Class B tube circuit.

identical to the tube version with the collectors attached to the two ends of the audio output transformer. Except for the fact that solid-state devices are used instead of the tubes, the servicing is the same.

CLASS B TAPPED SPEAKER (PUSH-PULL) 40

It is desirable, for economy reasons, to eliminate the output transformer. The savings are considerable in parts, labor, and keeping the size of the chassis small. A few ways have been devised to eliminate the output transformer.

One of them is with the use of a specially made speaker with a tap on it (Fig. 76). The tap is designed so that the impedance on either side of the tap is the same. Then the collectors of the two transistors are attached to either end of the speaker. The voltage source is attached to the tap. The speaker then acts as an autotransformer and becomes its own output transformer. Servicing is again essentially the same, but care must be taken not to operate the amplifier without the speaker. When that happens, the absence of a load on the transistor will destroy it.

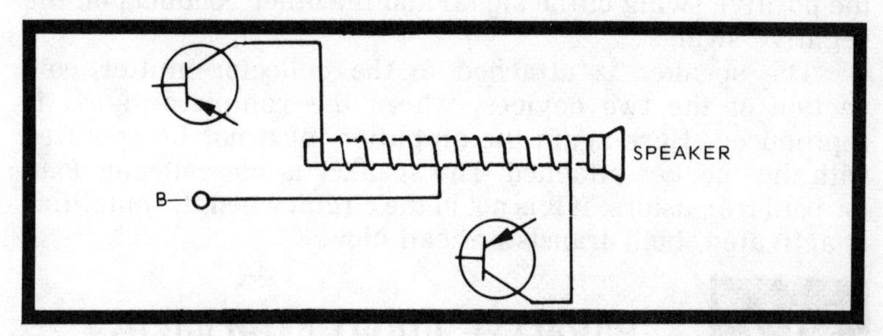

Fig. 76. An output transformer can be eliminated with a speaker designed to accept the signal directly.

CLASS B CASCODE (PUSH-PULL) 41

Another common transistor configuration without an output transformer is a pair of transistors stacked in cascode. Typically, two npn transistors are arranged one on top of the other (Fig. 77). The emitter of the top one is attached to the collector of the bottom one.

The input transformer has two secondaries. Each secondary feeds a different phase signal into each base-

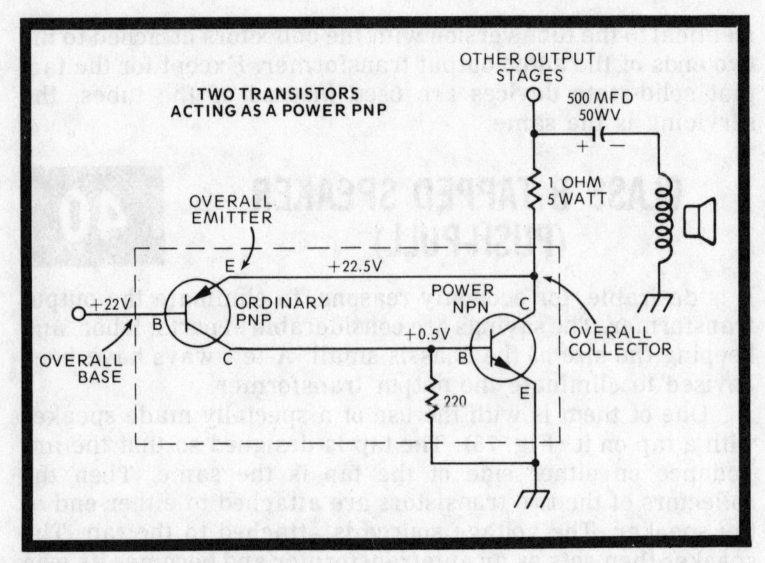

Fig. 77. A cascode circuit uses "stacked" transistors. In this case, the top E and bottom C are connected.

emitter junction. That way one of the transistors conducts on the positive swing of the signal and the other conducts on the negative swing.

The speaker is attached to the collector-emitter connection of the two devices, where the complete signal is reproduced. Here again the amplifier must not be operated with the speaker attached. The speaker is the collector load for both transistors. If it is not in the circuit when the amplifier is activated, both transistors can blow.

42 PARALLEL AUDIO AMPLIFIERS

Audio output transistors can be connected in parallel. In Fig. 78 the signal is applied to the base and emitter of a transistor. An audio output transformer is attached across the collector and extracts the amplified audio. Then one or more identical transistors are simply soldered onto the first one. All the bases, emitters, and collectors are connected together.

The efficiency of this circuit is low, but there is an additive effect to the gain. Two transistors hooked up in such a manner provide more output than one. Three provide more than two, and so forth. But why bother? Why not just use the vastly more efficient push-pull, you ask? It seems that some manufacturers have found that if they advertise a 20-transistor radio for the same price as a 6-transistor radio, it will sell quicker. Never mind that the 6-transistor might have gain. The large numbers sell radios. The 20-transistor-model manufacturer may have used cheap devices and inexpensive production, then simply soldered a number of transistors into the spot that one should be in. There is more gain, but not much more, so the advertisement of the large number of transistors is not false.

Servicing these circuits can be tricky. If one of the transistors shorts, the only way you can determine which one it is, is to remove all the parallel units.

Other Classes of Amplifiers

The preceding discussions of Class A and Class B amplifiers arouses interest in the other amplifier classes, such as Class AB, Class AB_1, Class AB_2, and Class C. The class of an amplifier, whether it is tube, transistor, FET, or any other amplification device, is determined by the number of degrees in the total of 360 degrees of signal the amplifier is designed to pass.

Signal is typified by a sine wave, which is actually a circle that is drawn through time. The sine wave starts at zero

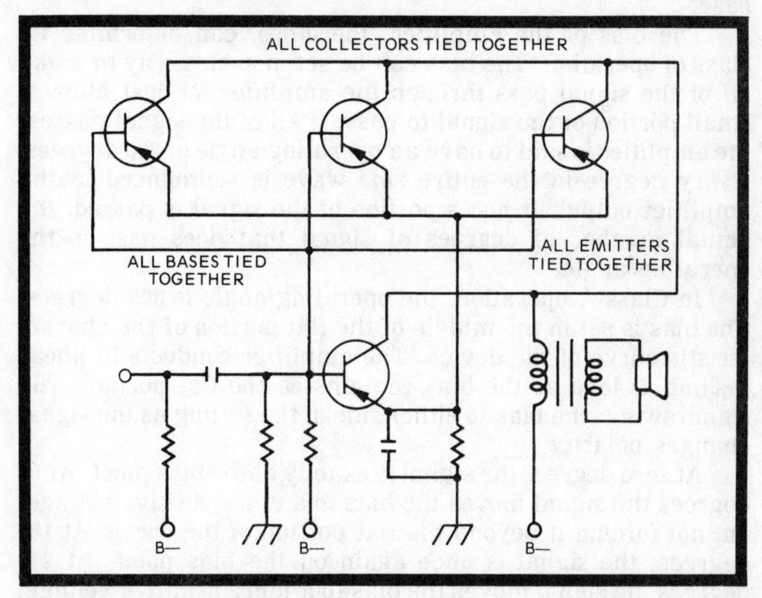

Fig. 78. A bit more gain can be produced by connecting a group of transistors together in parallel.

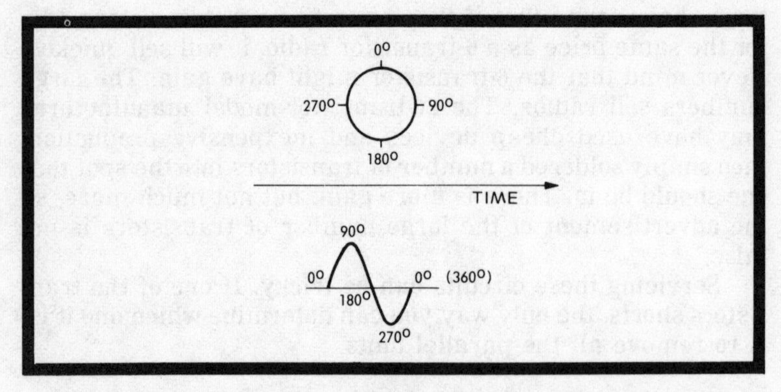

Fig. 79. When a circle is traced through time, a sine wave is formed.

degrees, rotates clockwise to 90 degrees, which is the highest positive amplitude, keeps rotating to 180 degrees and then down to 270 degrees, the lowest negative amplitude. Then it returns to zero degrees again (Fig. 79). Since it is rotating on a time base, the complete rotation, instead of appearing as a circle, becomes the familiar sine wave. Always remember, though, it is a circle. Any CW (continuous wave) note is a circle that describes a sine wave as it passes along its time base.

The bias of the amplifier, therefore, can determine its class of operation. The bias can be set in such a way to make all of the signal pass through the amplifier, or just allow a small portion of the signal to pass. If all of the signal passes, the amplifier is said to have an operating angle of 360 degrees. Every degree in the entire sine wave is reproduced in the amplifier output. If just a portion of the signal is passed, the actual number of degrees of signal that does pass is the operating angle.

In Class A operation, the operating angle is 360 degrees. The bias is set in the middle of the flat portion of the characteristic curve of the device. The amplifier conducts in linear fashion as long as the bias remains on the flat portion. The signal swings the bias to either side of the setting as the signal changes polarity.

At zero degrees the signal is exactly at the bias point. At 90 degrees the signal moves the bias to a more positive voltage, but not forcing it beyond the flat portion of the curve. At 180 degrees, the signal is once again on the bias point. At 270 degrees the signal moves the bias to a more negative voltage, but not beyond the flat part of the curve. At zero again the signal returns to the bias point. During all 360 degrees of the continuous wave's rotation, the device conducts and amplifies the input signal.

During Class B operation, each device is said to have an operating angle of 180 degrees. The design is such that if two amplifiers are arranged to operate out of phase with each other, the two will be able to reproduce the entire 360 degrees of the CW. One will deliver the signal between zero and 180 degrees and the other between 181 to 360 degrees to a mutual output.

The bias on each one is set at cutoff. With no signal applied, both devices are turned off. As signal appears at the input, the following happens: As the signal passes zero degrees one of the devices turns on. The signal forces the bias away from cutoff. As the signal passes through 90 degrees the device conducts the heaviest, since that is the highest positive amplitude. Then, as the signal reaches 180 degrees the bias is returned to cutoff.

However, at 181 degrees the second device in the push-pull arrangement is brought away from cutoff. It turns on. At 270 degrees it conducts the heaviest, since that is maximum negative amplitude, and the device is set up out of phase with the first one. Then the signal reaches 360 degrees, the device turns off and the first one prepares to turn on again.

Distortion

There are many applications where it is not necessary to have a distortion-free output. For instance, in rf amplifiers used in transmitters. The important thing is the signal rf. It can be reproduced with an operating angle much less than 180 degrees. This is called Class C operation.

A big advantage in Class C operation is the plate or collector efficiency. One of the major reasons for losses in the plate or collector circuit is the heat built up during current flow. This type of heat generation is reduced if the amount of time current flows is reduced in the output circuit. When the operating angle is way below 180 degrees, the amount of time conduction occurs is much less than when the operating angle is 180 degrees or more. In Class C operation of an rf amplifier, the only desired result is the amplification of power of the rf signal. The distortion is unimportant.

Efficiency

The less time a device conducts, the better the efficiency. Also, the better the efficiency, the greater the distortion.

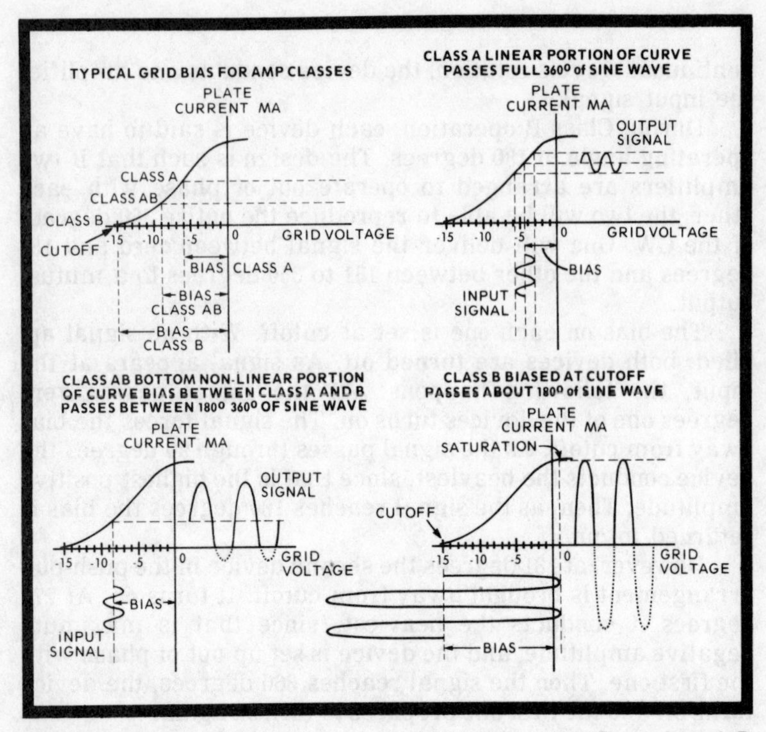

Fig. 80. A Class AB amplifier is biased between the Class A and B operating points.

Therefore, a Class A amplifier, which has an operating angle of 360 degrees, has poor efficiency and practically no distortion. A Class B amplifier, which has an operating angle of 180 degrees, has better efficiency but distorts all the singal between 180 and 360 degrees. A Class C amplifier, which has an operating angle of less than 180 degrees, has the best efficiency but distorts the signal the most.

The AB Classes

During the design of amplifiers, some applications require more efficiency than can be derived from Class A, yet cannot tolerate the large amount of distortion present in Class B. So a class called AB is used (Fig. 80). The bias is set somewhere between cutoff (Class B) and the center of the straight portion of the characteristic curve. This design provides an operating angle somewhere between 180 and 360 degrees. Class AB amplifiers are rarely used alone. They are typically used in push-pull designs, as are Class B.

The list of classes goes on. There is a Class AB_1, where the bias is set so that in a tube the grid never goes into the positive region and draws current. In a transistor the base never exceeds its normal bias and saturates. Then there is Class AB_2, where the bias is set so that in a tube the grid can go positive at some angles of the input signal. In a transistor the input signal does cause the transistor to conduct too much and saturate, distorting the signal during that part of the operating angle.

SIMPLE CODE PRACTICE OSCILLATOR 43

Every embryo ham radio operator starts out with a code practice oscillator. It is a good example of how an oscillator works and fails.

An oscillator is nothing more than an electronic circuit that turns itself on and off automatically once it is activated. The value of the circuit components determines the off-on switching frequency. In the oscillator shown in Fig. 81 the frequency is determined by the time constant of the resistor and capacitor. An npn transistor is connected directly, collector-to-base, to a pnp transistor. Then there is a 0.01 uF capacitor feeding back from the pnp collector to the npn base. In the npn base to ground is a 100K resistor. The output goes to a speaker from the pnp collector and the npn emitter. A battery with a key (off-on switch) is attached to the speaker.

When the key is pressed, the negative battery circuit is closed. This places an excess of electrons on the collector of the pnp (through the speaker) and the emitter of the npn. It also places an excess of electrons on the collector side of the capacitor. That repels electrons from the other side, thus

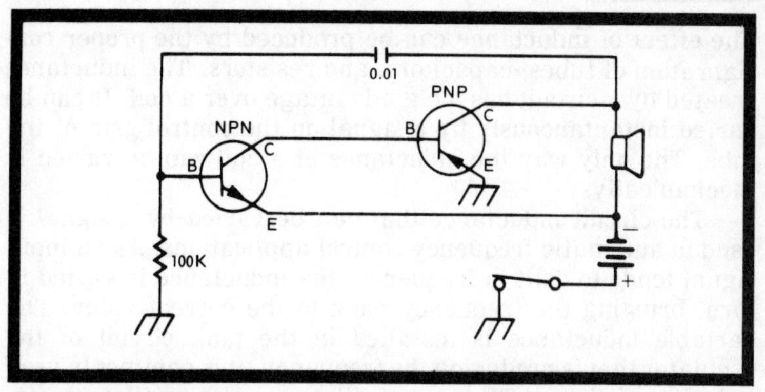

Fig. 81. A code practice oscillator is a simple oscillator circuit designed to operate somewhere in the audio frequency range.

charging the capacitor. The repulsed electrons start to leak off to ground through the 100K resistor. This causes a deficiency of electrons on the base of the npn. The deficiency is a positive charge on the base. There is an excess or a negative charge on the emitter from the battery. This constitutes forward bias across the EB junction and electrons start to flow from E to B.

The EB electron movement makes some electrons move from E to C. Since C of the npn and B of the pnp are connected together, electrons go to the pnp base, instantaneously providing an excess of electrons. Since E of the pnp is at ground, the charge at the base constitutes forward bias and electrons flow from B to E of the pnp. This causes some of the excess of electrons on C to also move toward E.

As electron flow starts on the pnp collector to ground, the excess of electrons on C turns instantaneously to a deficiency. This takes the charge away from the capacitor, which in turn changes the charge on B of the npn. The charge on B being lost causes the npn to turn off.

The effect of the npn turning off is coupled to the base of the pnp, which turns it off. This restores the battery voltage to the capacitor, recharging it and turning the npn on. This in turn makes the pnp conduct. The cycle continues as long as the key is held down. With the 0.01 uF capacitor and the 100K resistor, a 1000 Hz note comes out of the speaker. This is an RC-type oscillator and the frequency of the tone can be varied by changing the resistor or capacitor or both to other convenient values.

ELECTRONIC INDUCTANCE (REACTANCE TUBE)

The effect of inductance can be produced by the proper configuration of tubes, capacitors, and resistors. The inductance created by a circuit has a big advantage over a coil. It can be varied instantaneously by a signal on the control grid of the tube. The only way the inductance of a coil can be varied is mechanically.

The circuit inductance that can be varied by a signal is used in automatic frequency control applications. As an input signal tends to drift in frequency, the inductance is varied in turn, bringing the frequency back to the correct value. The variable inductance is installed in the tank circuit of the oscillator that is producing the frequency. It is commonly used to lock in radio and TV tuner oscillators, color oscillators, etc. Let's see how a circuit can imitate an inductance.

When a signal voltage is applied to a coil, a back voltage is generated which presents a resistance to the electrons as they try to enter the coil. Therefore, at the largest value of voltage, the resistance from the back voltage is the greatest and the least amount of current can get into the coil. Conversely, when the voltage passes through zero, little or no back voltage is generated and the resistance to the electrons is near zero, permitting electrons to enter in the greatest intensity. This boils down to the fact that the current lags the voltage about 90 degrees. That is the definition of an inductive reactance. The current lags the voltage by 90 degrees, or, to put it another way, the voltage leads the current by 90 degrees. If we can set up a circuit that can produce this effect, we can create a coil-less inductance. This circuit can be installed in the oscillator as an actual inductance that will vary according to its input signal which comes from the oscillator.

A pentode tube with resistors and capacitors is used (Fig. 82). The oscillator is connected across the plate and cathode. The oscillator signal is applied to the plate circuit through a coupling capacitor. In series with the plate capacitor is a second capacitor, a resistor, and the grid-leak capacitor. The signal flows in this series circuit to ground. The resistor is a high value. This makes the series circuit mostly resistive and not reactive. That means no phase change takes place and the current and voltage from the signal are still traveling together in phase till they reach the grid-leak capacitor.

In the grid-leak capacitor, before it is charged, the signal meets no resistance. The current flows in full force as the voltage in the capacitor passes through zero. Then as the capacitor fills up and the voltage in the capacitor rises, the

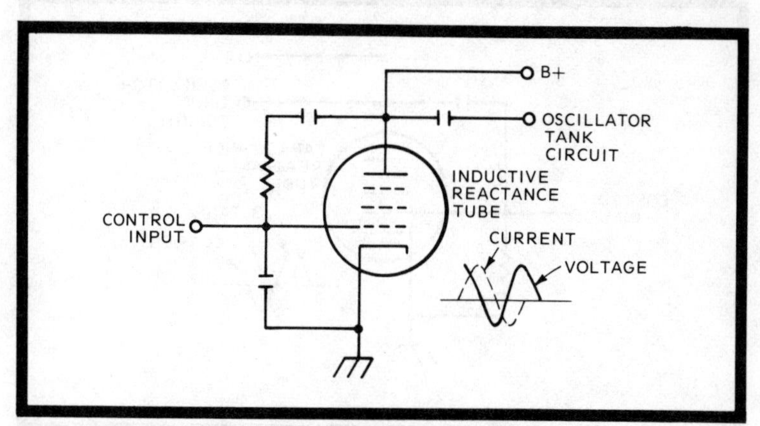

Fig. 82. An amplifier circuit can be made to appear as an inductance when its output current lags 90 degrees behind the voltage.

current turns off. The capacitor makes the voltage lag the current by near 90 degrees.

The control grid is tied to one end of the grid-leak capacitor. In a tube the plate current is in phase with the grid voltage. Since the incoming voltage is lagging, the plate current will also lag. This plate current is injected into the oscillator tank. The lagging current represents a certain amount of inductance.

The amount of inductance can be controlled by the amount of plate current that is injected into the oscillator tank. This is a function of the bias on the tube. If more plate current is allowed to flow by making the bias more positive, a low amount of inductance is produced. If less plate current is allowed to flow, by raising the grid bias, a larger inductance is produced.

45 ELECTRONIC CAPACITANCE

In a similar way, an actual capacitance can be produced, a capacitance that can be varied by an incoming signal and used to control the frequency of an oscillator. The circuit is a bit simpler (Fig. 83). A grid-leak resistor is used and a capacitor is connected between the control grid and plate. The oscillator is again connected across the plate and cathode. This time the result is that the plate current ends up leading the applied voltage by near 90 degrees. This causes a certain amount of capacitance to be injected into the oscillator tank circuit.

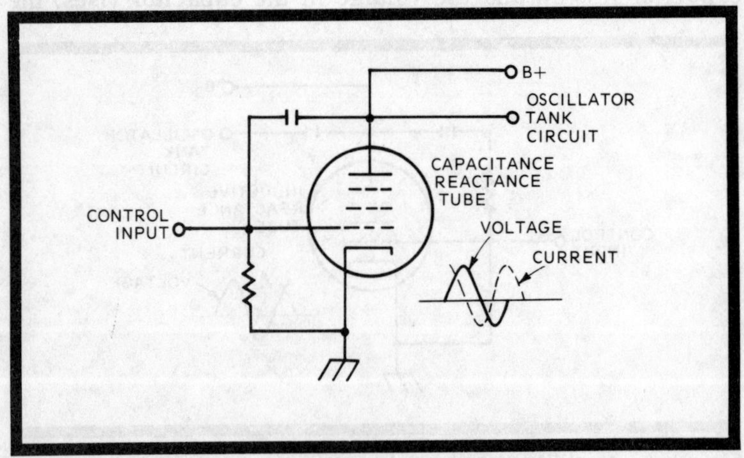

Fig. 83. An amplifier circuit can be made to appear as a capacitance when its output current leads the voltage by 90 degrees.

If the bias on the tube causes more plate current, there will be more capacitance applied to the oscillator tank. If the bias on the tube causes less plate current, there will be less capacitance. More capacitance lowers the oscillator frequency, while less capacitance makes the oscillator increase in frequency.

46 NEUTRALIZING CIRCUIT

In all tubes and transistors, there is a certain amount of capacity between the elements. In tubes, the elements are electrodes separated by a dielectric, a vacuum. In transistors the elements are the pn junctions that have a certain capacity according to the width of the junction, regulated by the amount of bias. Therefore, any ac input has two paths available. It can travel through the device as it modulates electron flow, or it can travel through the capacitance between the tube or transistor elements.

The signal is primarily thought of as traveling through the device via the electron flow. The capacitance path is usually unwanted and is even referred to as parasitic capacity. However, it is present and must be considered in a circuit. At low frequencies, the capacity has little effect and can be considered negligible, but as the frequency is increased, it becomes a factor.

In the primary path through an amplifier, the ac enters the grid or base and leaves the device 180 degrees out of phase with its input. The secondary path through the capacitance has the ac entering the grid in the same way but leaving the device with the same phase it had as it entered. In the output, then, there are two signals 180 degrees out of phase with each other. The primary signal is amplified and the secondary one is not. The secondary cancels its amplitude out of the primary. This leaves gain, but the losses can be significant.

It is desirable to get rid of that signal that passes through the capacitance of the device. It's done by placing into the output a third signal, equal in amplitude but opposite in phase to the parasite. A capacitor is installed from the input to the input of the next stage, which is across a transformer (Fig. 84). Thus, a third path for the signal has been established. After the two signals from the first two paths pass through the transformer they encounter the third signal.

The number one signal is the amplified desired signal. The number two signal is the capacity-coupled parasite. The number three signal is the capacity-coupled neutralizing

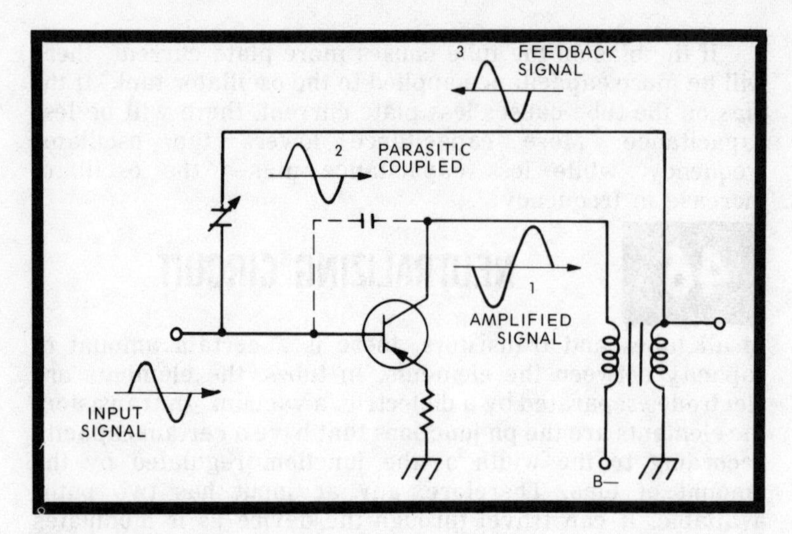

Fig. 84. The grid-plate interelectrode capacitance of an amplifier can be neutralized with a feedback signal.

signal. The number one and number three signals are in phase and the number two is 180 degrees out. Number three tends to cancel number two and reinforce number one. The net result: The parasitic signal is **neutralized** out.

No one really knows where the neutralizing takes place. Some sources say the number three signal is being coupled back to the original input and canceling three. It doesn't really matter, the important thing is the cancellation. The coupling capacitor is usually variable and is adjusted until the neutralization process is as close to perfection as possible. Neutralizing is most common in rf and i-f stages since frequencies are so high.

47 SIMPLE VOLTAGE STEPUP

The power supply of any electronic gear has the job of producing a dc voltage. If a positive potential is needed, the supply attracts electrons from the gear and causes a deficiency of electrons at key points such as the plates or collectors of npn transistors. If a negative potential is needed, such as for the collectors of pnp transistors, the power supply can add electrons to the circuit and cause an excess of electrons to appear on the collectors.

The voltage levels needed vary. Since the ac source typically is 117V, voltage values have to be manipulated. Various circuit configurations deliver various voltages.

Suppose you have a one-tube circuit that needs 160V. A simple half-wave rectifier will deliver it from the 117V source. The only requirement is that the tube draws very little plate current. How can 160V be derived from 117V? The peak voltage of the 117V house current is actually 1.4 x 117 or 163.5V (rms). If you draw small amounts of plate current, about 160V will appear on the plus side of the filter capacitor attached to the cathode of the rectifier (Fig. 85). If you start drawing any real amount of current out of the plate circuit, though, the voltage will drop quickly. However, this peak-voltage gimmick is quite useful and is found in many applications where little or no current is needed, just voltage.

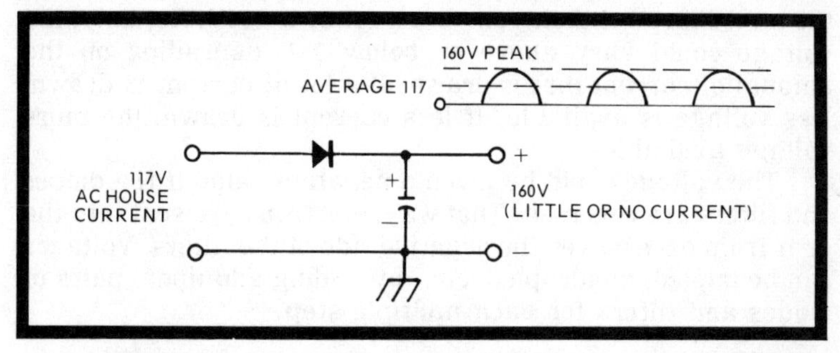

Fig. 85. When little or no current is drawn from a power supply, the peak line voltage is produced.

SIMPLE VOLTAGE DOUBLER 48

Twice as much voltage as is supplied by the 117V source can be produced in a simple voltage-doubler circuit. It is almost identical to the simple voltage stepup, except another filter and diode are installed (Fig. 86). When the gear is turned on, the first half of the sine wave causes the first filter to charge as electrons flow through the first diode. Then, as the second half of the sine wave appears, the first diode shuts off and the second diode conducts, charging the second filter.

The separate charges on the two filters are maintained by the incoming sine wave. Each filter is charged with the 117V potential. Electrons are attracted from the gear by both potentials. C1 attracts electrons through D1. D1 is forward biased and has little or no resistance. C2 attracts electrons

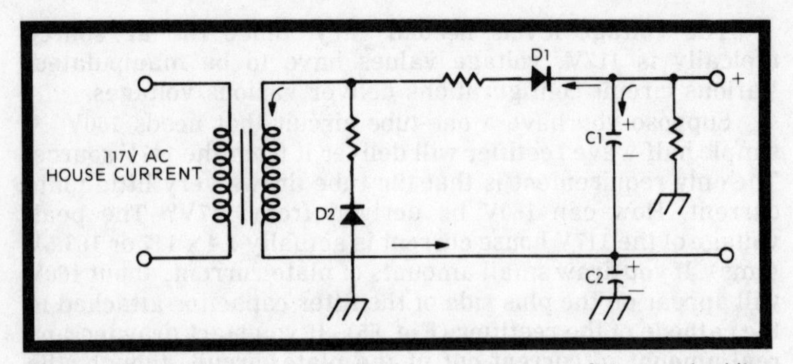

Fig. 86. Line voltage can be doubled by charging two filters and adding their charges together.

directly from the gear. Each filter is charged with 117V and the total charge is their sum or 2 x 117 or is 234V. Actually, the voltage could vary above or below 234, depending on the amount of current that is drawn. If a lot of current is drawn, less voltage is available. If less current is drawn, the more voltage available.

The voltage could be given a negative value if the diodes and filters are reversed. That way, electrons are sent into the gear from excesses on the negative side of the filters. Voltages can be tripled, quadrupled, etc., by adding additional pairs of diodes and filters for each multiple step.

49 ALTERNATE VOLTAGE DOUBLER

Another voltage doubler circuit shows the two filters in series, but that is a confusing way to view it (Fig. 87). Actually, the bottom filter is in the same position as in the previous circuit, except for the fact that is on the other side of the transformer. I point this out because as a servicer, you'll encounter this circuit frequently and it's best not to think of the two filters in series.

When the gear is turned on, during one half of the sine wave electrons are drawn out of the gear, pass through the top diode, through the bottom diode to ground. As a result, the top filter charges. As the second half of the sine wave passes, neither diode can conduct, but electrons are drawn from ground into the negative side of that bottom filter. The negative charge forces electrons from the positive side, causing that filter to charge. The two charges then stay on the filters. The deficiency of electrons on the top filter and the

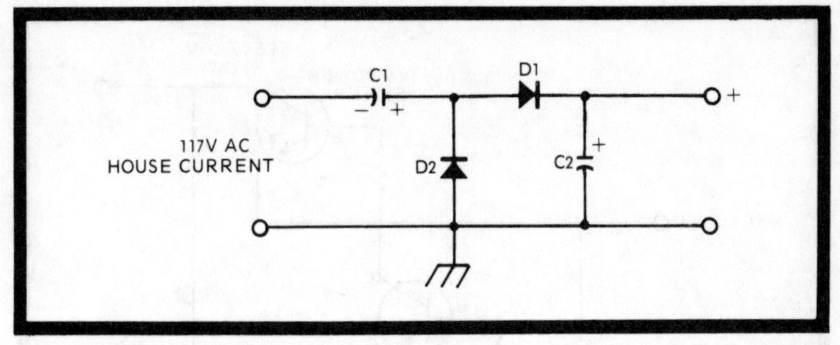

Fig. 87. Voltage can be doubled by putting filters on each end of a power transformer secondary as shown.

deficiency of electrons on the bottom add together through the source and the top diode.

Depending on the amount of current drawn, the voltage developed in this supply can vary. If little or no current is drawn, the voltage can approach twice the peak value of the 117V, that is 117 x 1.4 rms x 2 or 326V. As more and more current is drawn, the voltage will drop accordingly. The resistors in series with the diodes are there to limit the initial surges of current as the filters charge.

ACTIVE FILTERING CIRCUITS 50

In preceding discussions the role of the filter capacitor and filter choke was touched upon. Filtering can also be performed with a transistor (Fig. 88). The transistor is used in conjunction with capacitors and chokes. In high-voltage circuits, where B-plus values are in the hundreds of volts, filtering is easy. Capacitors and chokes perform the job. In solid-state circuits where the voltage is down around 10V, larger filters are needed so they can maintain the charge between pulses. While 20 uF and 50 uF capacitors hold a charge between pulses if the voltage is in the hundreds, a 1000 uF capacitor is needed to do the same job at 10V.

Even with the high-value filters, enough of the filter charge can leak off to cause disruptive ripple in the B-plus. In order to cancel out the ripple, an active transistor filter is used. They are usually large transistors and operate in a noise-cancellation configuration. The typical hookup shown in Fig. 88 has an active filter and filter driver transistors. As electrons are attracted to the power supply by the B-plus potential, they find two paths: One to the collector of the pnp active filter and, two, to the emitter of the filter driver.

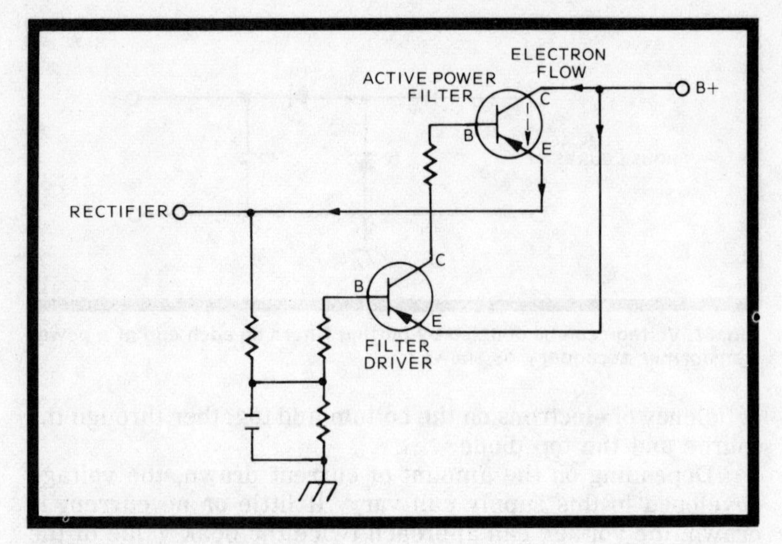

Fig. 88. The filter driver feeds ripple to the power filter 180 degrees out of phase, which cancels the ripple.

Electrons flow into the filter collector and out the emitter to the rectifier. Most of the electrons pass that way. Since the current flow is directly through the active filter, it must be a large transistor in order to endure the heavy flow. In the electron flow is contained any ripple that might be left from the inability of the filter capacitors to hold their charge during the time between pulses.

Meanwhile, the filter driver collector is attached to the base of the active filter. The base of the driver is connected to the B-plus line going to the rectifier. It takes a steady sample of the B-plus including the ripple. The ripple modulates the driver collector current. The driver then amplifies and inverts the ripple. Since its collector is attached to the base of the active filter, this inverted ripple modulates the B-plus passing through the active filter. The inverted ripple, 180 degrees out of phase with the ripple in the B-plus, cancels the B-plus ripple.

The base-to-ground voltage divider of the driver is usually a potentiometer. Thus, the size of the amplified ripple can be adjusted. Proper adjustment eliminates the unwanted B-plus ripple completely.

Active Filter Circuit Failure

These transistors are subject to failure due to the fact that the B-plus must pass through the active filter. The symptoms of failure are identical to the symptoms of defective filter capacitors. Most often the transistors short. When they do, the B-plus still passes but no filtering takes place. Occasionally, the filters open. Spraying them with a freezing agent could heal them temporarily, revealing the defect. These circuits are becoming very common.

RECTIFIER PROTECTION CIRCUIT 51

It is quite common that after a lightning storm a TV or radio won't work. The servicer discovers the solid-state rectifiers in the low-voltage supply are blown. They can be shorted open or physically blown apart.

Why does that happen? During the storm, lightning hits the power line. Lightning voltages are in the millions of volts. However, the lightning doesn't usually hit the wire, but the wooden pole. The high resistance of the pole stops most of the current and causes a large voltage drop. Also, the lightning hits for an instant. What actually gets into the line is a transient voltage spike. By the time it rides into a plugged-in appliance, it has a value of a few hundred or a few thousand volts. But this is enough to jump the open off-on switch and do a job on the rectifier, which is not rated to handle such a voltage. What can be done? A permanent cure is easy.

Since the spike is so short in duration, it has a high frequency. All that is needed is a bypass capacitor across the input line that has a low reactance to a high frequency and a high reactance to a low frequency, such as the 60 Hz line frequency. A 0.01 uF capacitor (or thereabouts) has about 330K reactance to 60 Hz and practically no reactance to the transients. Any transient spike that enters will be short-circuited right to ground before it can get to the rectifiers (Fig. 89). The capacitor will also filter out any high-frequency interference that might enter due to other appliance-generated interference.

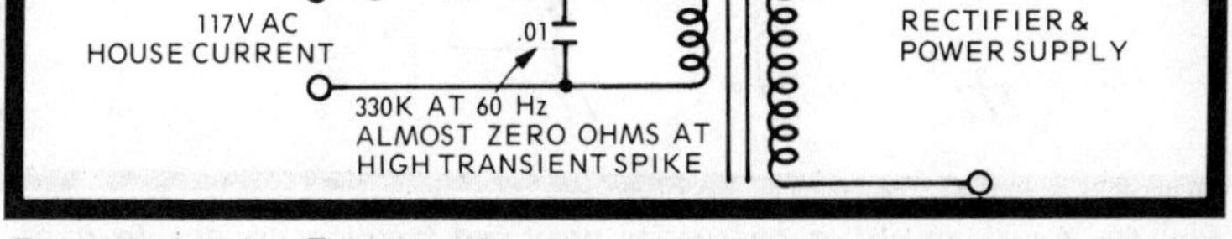

Fig. 89. A 0.01 uF capacitor across the line input will bypass high-frequency transient voltage spikes.

52 RF AMPLIFIER RECEIVER CIRCUIT

An rf amplifier's main function is to pick one desired frequency out of the myriad of transmitted signals. The important part of the rf amplifier is its tuned input and output. The input has to have a particular impedance, which should match the antenna because the antenna feeds the amplifier. In fact, you could say that the antenna and input components are all part of the rf input.

The input is tuned by variable capacitive or inductive components over the designed range of the receiver (Fig. 90). The tuned components couple the signal into the amplifying device. The device could be a tube, transistor, or FET. The rf amplifier then does the following jobs.

One, since it is located between the antenna and the following stages, the rf stage stops any interfering signals from getting into the receiver by its selective tuning. These are signals like electrical interference, image, and i-f frequencies. If the rf amplifier was not there, these other frequencies would be able to make it through the other stages.

The rf amplifier must have an excellent signal-to-noise ratio. Since it is the first stage in the receiver, its output, if clear, will appear clear in the speaker or on the crt. If there is

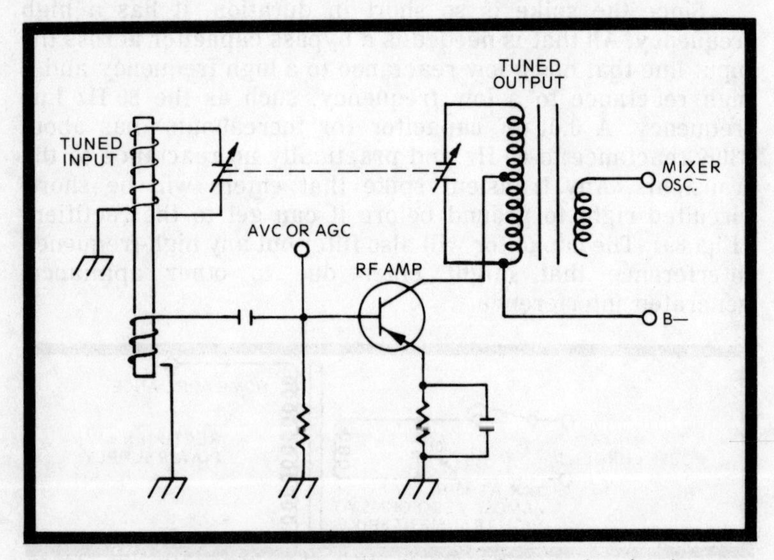

Fig. 90. An rf amplifier passes its designed frequencies due to gang-tuned input and output circuits.

some noise in the signal, then the noise will be amplified along with the signal and appear many times worse in the output.

Rf amplifiers have gain, but as mentioned the gain is not that important. The gain simply has to be of a figure where the signal-to-noise ratio is good. Further on in the i-f, video, or audio stages, more gain can be accomplished. Only enough gain is needed to have the rf output as noise-free as possible. It is found that a gain figure between 50 and 100 is usually quite adequate for the rf amplifier.

Also, the amplifying device should have a low level of internal noise. In tubes, triodes have low noise, tetrodes medium noise, and pentodes have a high noise characteristic. The designer picks the device, of course. FETs have been found to have excellent low-noise characteristics, while transistors have a somewhat higher level.

An rf amplifier must be selective. The input tuning must be as sharp or as broad as is necessary to pluck the desired frequency out of the air in its entirety. An AM radio has very selective rf tuning since a plus or minus 5 kHz bandwidth is all that is needed to receive one station. A TV rf amplifier, on the other hand, has to be able to receive a band about 4 or 5 MHz wide in order to get all the signal. An rf amplifier must produce the same amount of gain throughout its tuning range. Also, it should produce the same amount of gain across the bandwidth of the selected station. This linear characteristic is important.

Lastly, the rf amplifier must provide good isolation between the local oscillator and the antenna. If it doesn't, the oscillator and antenna become a transmitter, radiating signals throughout the area.

RF Circuit Tests

An rf amplifier is best tested with a signal generator. That way a specific signal frequency can be injected and the amplifier operation observed. Any decrease in gain or overloading can be seen. If the gain is lost altogether, that is obvious, too. The generator is attached to the antenna and a low-frequency signal turned on. In an AM radio that would be 55 kHz; in an FM radio, 88 MHz; and in a TV, Channel 2 or around 56 MHz. The signal is modulated with a 400 Hz note. The note should be heard from the speaker or in a TV black horizontal bars should appear on the screen.

If the note or the black bars are missing, weak, or overloaded, the same signal is then injected into the input of the next stage. Should the resultant signal then appear normally, the trouble has been isolated to the rf amplifier. Then dc voltage and resistance tests are made to pinpoint the actual component.

RF Amplifier Overload Complications

When the receiver is overloading and you isolate the trouble to the rf stage, you must disable the avc or agc circuit before concluding definitely that the rf stage is at fault. In tube receivers, a bias box can be used as a substitute for the agc voltage. Attach the bias box with the proper bias voltage to the rf input from the agc circuit. Then adjust the bias voltage around its prescribed value. Should the overloading clear, then the rf amplifier is not in trouble; the agc circuit is. (See the agc discussion further on in the book.) In solid-state receivers, the bias box technique is not applicable. In that case, the agc circuit must be considered a suspect and checked along with the rf amplifier.

53 DIODE MIXER CIRCUITS

The superheterodyne receiver, which includes all TVs and most radios on the market, is so called because it converts any incoming rf to a lower frequency. It doesn't matter which end of the dial is tuned, this lower intermediate frequency remains the same. In the tuner, along with the rf amplifier, is a local oscillator that tracks with the rf tuning. The rf and oscillator outputs are combined in a mixer, and the difference frequency is the i-f.

In the UHF range an amplifier tube is not very efficient in mixing the rf and oscillator outputs. A diode with an extremely high forward bias does it better. It's called a hot-carrier diode since the forward bias is about 20V. This type of diode has a low noise figure and frequency response in the UHF range holds steady. The rf and oscillator signals are coupled into the diode by perfectly matched coaxial cable (Fig. 91). There is little or no loss that way. The only loss is the energy it takes to heterodyne the two frequencies, called conversion loss.

A good rf amplifier must be used ahead of the diode mixer and a good i-f stage afterwards, since there is no gain, just loss, in a diode. In order to keep the losses at a minimum, four diodes in a bridge circuit can be used instead of just one. The balance of the bridge gives a 72-ohm input and a 72-ohm output. This is convenient since the coupling can be accomplished with pieces of 72-ohm coaxial cable.

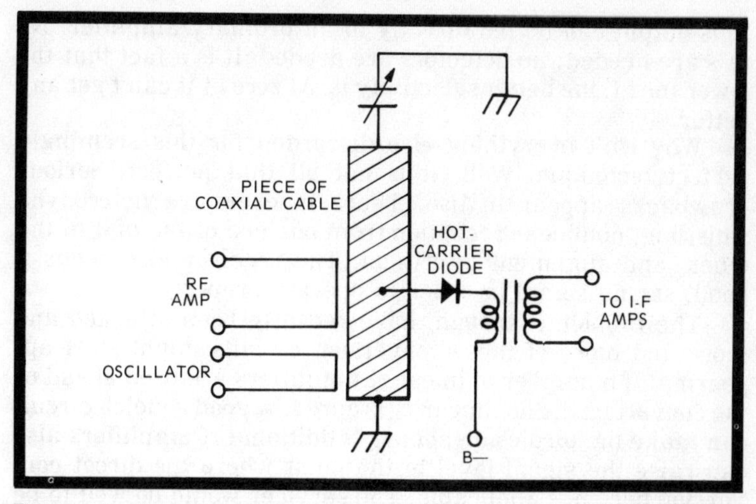

Fig. 91. A hot-carrier diode is so named because it has an abnormally high forward bias. These diodes are frequently used as mixers of high frequencies.

DIRECT CONVERSION CIRCUIT 54

Another heterodyne technique is called direct conversion. The local oscillator is set to run at exactly the same frequency as the incoming rf signal. Then they are both injected into a mixer circuit (Fig. 92). The difference frequency is zero, since there is no difference. What good is that? The modulation around the zero point is automatically detected! The zero i-f is already detected. The rf and oscillator cancel each other out.

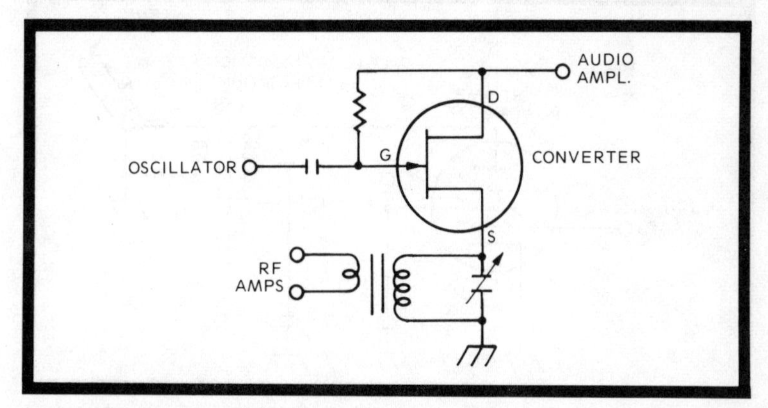

Fig. 92. Direct conversion can be accomplished by beating a carrier wave with a signal of the same frequency.

This output can be fed directly to an ordinary amplifier. No i-f's are needed; no detectors are needed. It is a fact that the lower the i-f, the better selectivity is. At zero i-f it can't get any better.

Why isn't everything else discarded for this seemingly perfect technique. Well, it is not all that perfect. Serious drawbacks appear in use. There is excessive heterodyne whistling, nonlinear reception from one end of the dial to the other, and alignment is critical. The receiver also needs a good, strong signal in order to operate properly.

The problems, though, have recently been attacked and more and more of these conversion circuits might start appearing. If a number of linear rf amplifiers are used ahead of the converter, the nonlinearity is cured. A good squelch circuit can make the birdies disappear. Additional rf amplifiers also can raise the signal level to the point where the direct conversion becomes applicable. The servicer would do well to be aware of these circuits.

55 TRANSMITTER RF AMPLIFIER TUBE CIRCUIT

While the rf amplifier is the first stage in a receiver, it is the last stage in a transmitter. In a receiver it is a voltage amplifier; in a transmitter it is a power amplifier. The power output is needed to produce a heavy current in the antenna. The more electrons moving back and forth in the antenna at the radio frequency, the greater the electromagnetic field radiated from the antenna.

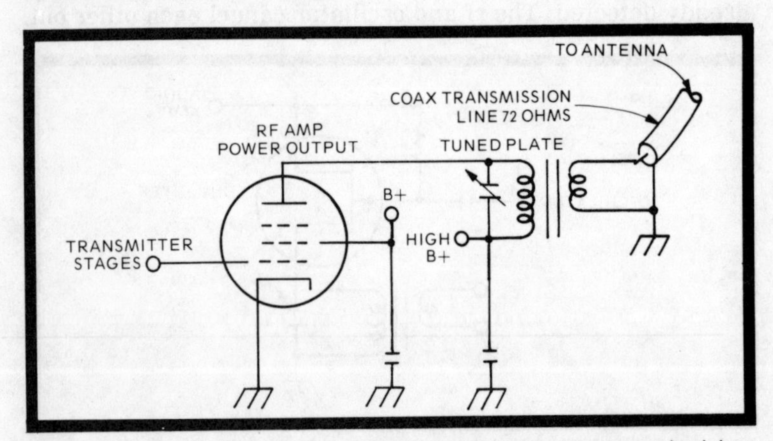

Fig. 93. An rf amplifier in a transmitter produces a power output to a tuned plate circuit.

The output of the rf amplifier must be coupled inductively into the antenna load. The primary of the output transformer is a tank circuit that resonates at the transmitted rf (Fig. 93). The secondary is a stepdown winding that matches the impedance of the large output tube to the lower impedance of the antenna. The antenna is typically a 72-ohm half-wave dipole.

TRANSMITTER RF AMPLIFIER TRANSISTOR CIRCUIT 56

In transistor rf output stage, the transistor acts exactly like the tube except for the impedances involved. The transistor has an output impedance of about 10 ohms. A stepup impedance must be used to couple the power to the 72-ohm antenna. A transformer can't be used because the primary of the transformer acting as a resonant tank circuit would generate a lot of harmonics.

The matching network used instead is a number of components and coils tuned to the operating frequency (Fig. 94). It's called an L network which is an inductance in series with three tunable capacitors. Correct alignment of the capacitors matches the transistor impedance to the antenna load.

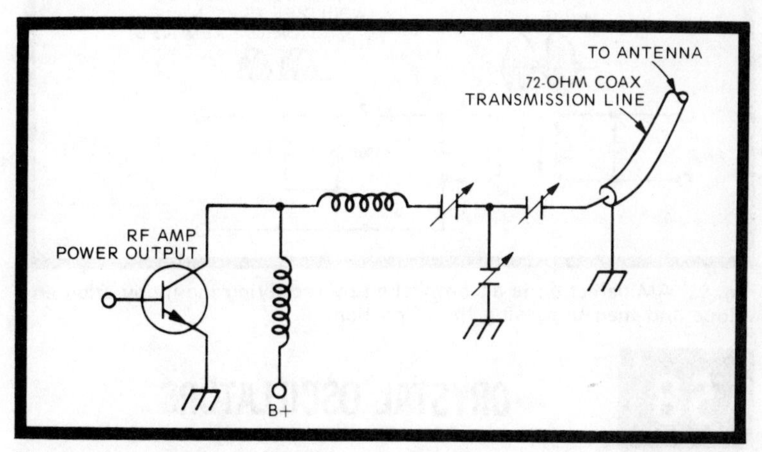

Fig. 94. A transistor rf amplifier needs a tuned output for best efficiency.

TIME-CONSTANT AM DETECTION 57

A diode performs rectification as the first step in the detection process. The signal is injected into the detector circuit and is rectified by the diode. The signal is typically an i-f amplitude modulated by intelligence, like audio. The rectified signal is developed across a load resistor in the cathode.

The actual "detecting" is done by a bypass capacitor. In an AM radio the capacitor is typically around 100 pF with a load resistor of about a half a megohm (Fig. 95). This time constant provides a bypass to the i-f but has little effect on the audio. The audio is developed across the load resistor.

As the electrons travel through the resistor and then from cathode to anode at the peak of the i-f voltage, the capacitor charges up to this value. Since the charge is maintained by each successive i-f voltage peak, the capacitor is a filter to the i-f. The modulation voltage peaks are much slower, so this change is not filtered and is the resultant waveform developed across the load resistor.

When testing for detector problems, an AM receiver circuit should have a time constant somewhere around 100 microseconds, although a wide variation is possible. Try to replace the resistor and capacitor, when necessary, with exact values; otherwise, you might change the time constant and upset the detection process.

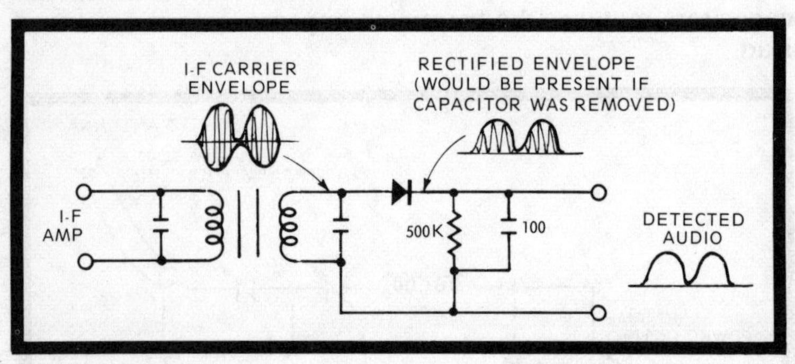

Fig. 95. AM detection is accomplished by rectifying the modulation envelope and then bypassing the rf portion.

58 CRYSTAL OSCILLATORS

It has been found that if a piece of quartz crystal is squeezed, a voltage is developed across the opposite faces of the crystal. This is called the "piezoelectric effect." The phenomenon works the other way, too. If a voltage is impressed across the crystal, it will physically squeeze itself or expand. When the voltage is ac, the crystal will vibrate. According to its size and shape, a crystal has a natural resonant frequency. When the ac applied equals the crystal resonant frequency, vibration is maximum. Also, a high quality crystal can maintain that frequency without variation. It is a stable oscillation device.

The thickness of the crystal is the frequency-determining factor. The length and width of the crystal has little effect. In order to attach wires to a crystal, two metal plates act as a holder, with the wires attached to the holder. A crystal and its holder has capacitance, inductance and resistance. In other words, it is a series resonant circuit.

Since a crystal is a tuned circuit, it can be put into the feedback circuit of an amplifier and start oscillating (Fig. 96). This turns the amplifier on and off as the voltage output of the crystal is applied to the control grid, gate, or base of the amplifier. The circuit will oscillate at the resonant frequency of the crystal. The circuit works the same whether a tube, FET, transistor, or integrated circuit is used. There are dual-gate ICs available expressly for the purpose of building a simple crystal oscillator.

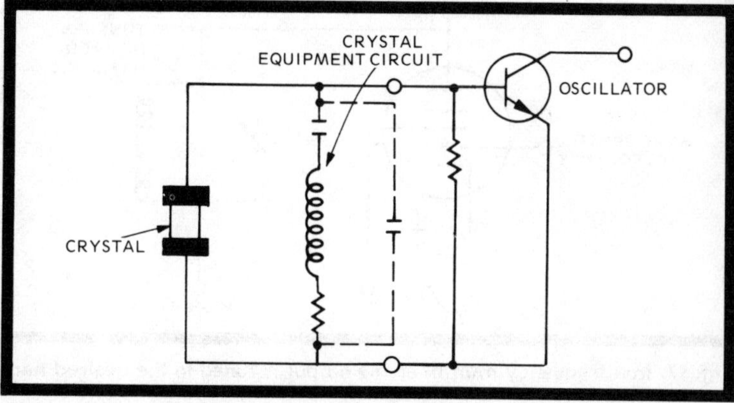

Fig. 96. In a crystal-controlled oscillator the crystal acts like a parallel and series tank circuit with some resistance.

FREQUENCY MULTIPLIER 59

The frequency of a crystal oscillator is determined by the thickness of the crystal. The thinner the crystal, the higher the resonant frequency. As the frequency goes up into the megahertz range, the crystal becomes extremely thin. At 10 MHz and above, the value of the crystal becomes doubtful because it is so thin. As a result, it is more practical to use the stable crystal to produce a lower frequency and then multiply the output to whatever frequency is required.

The multiplication process is enhanced by generating strong harmonics of the original frequency. If the amplifier is biased so that only a tiny portion of the input signal turns on the device, Class C operation is achieved. The output of a Class C amplifier contains just a few degrees of the sine-wave input and this high pulse output is rich in harmonics. Harmonics as high as the 50th are strong. Therefore, if you take a crystal tuned to 1 MHz, the 50th harmonic is 50 MHz with all the other harmonics in between available.

Frequency multiplication is valuable in test equipment like signal generators, where a means of checking and calibrating frequencies is a necessity. A variable capacitor control on the panel can tune the output to select the desired harmonic of the oscillator frequency (Fig. 97). The harmonic content of the output is determined by the amplifier's operating angle. The smaller the angle, the more and stronger the harmonic generation.

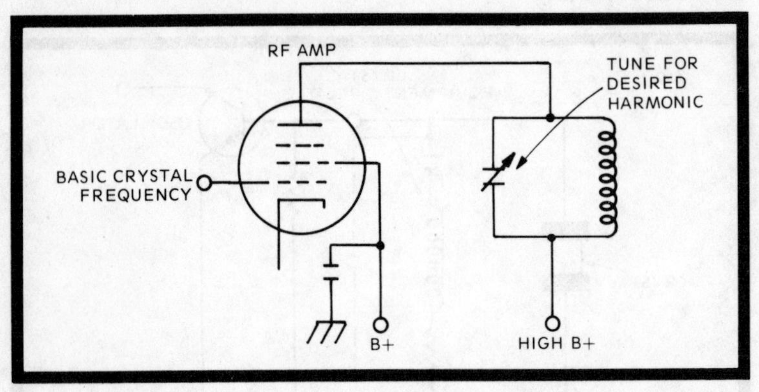

Fig. 97. In a frequency multiplier the output is tuned to the desired harmonic of the input.

60 FREQUENCY ABSORPTION CIRCUIT

A frequency absorption device is an instrument that is insensitive to all frequencies except the one it is tuned to respond to. The circuit consists of an absorption coil wound on top of another coil that is in series with a variable capacitor (Fig. 98). The second coil and capacitor are used to tune the absorption circuit. The absorption coil is in series with the cathode of a diode, which is in series with a microammeter. The diode acts as a rectifier. A filter capacitor from the diode to ground provides the microammeter with a steady dc. There

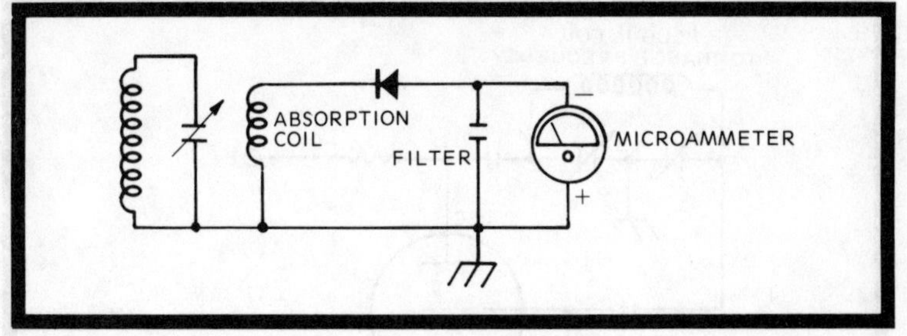

Fig. 98. Frequency can be measured with an absorption type circuit.

is no power supply in the meter. All the energy that activiates the unit comes from the outside rf.

The absorption coil is placed in an rf field. The electromagnetic energy of the field causes electrons to move in the absorption coil. The variable capacitor tunes the LC combination to the frequency of the surrounding field. At that point, a maximum amount of electrons move. The ac is rectified in the diode and charges the filter. The discharge from the filter passes into the microammeter and then to ground. The movement of the needle indicates the strength of the rf field. A strong field causes a high charge on the filter and a weak field causes a low charge. The dc from the filter is exactly proportional to the strength of the field.

DIP METER CIRCUIT 61

A frequency absorption meter is often confused with the dip meter, since they both measure frequency. They are actually exact opposites of each other. The absorption meter is powerless; it absorbs energy from an rf radiation field. The dip meter is an oscillator circuit (Fig. 99), fully powered, that generates a field around a plug-in coil. (Most dip meters have several coils to offer a wide range of measurements.) It is used to measure the resonant frequency of tuned circuits, mostly coils, which are not in an active circuit. (An absorption meter measures the frequency a coil is radiating while it is turned on.)

When the dip meter's oscillator is turned on, a small rf field is generated around the dip meter plug-in output coil. Then the dip meter is placed near the tuned circuit in question. At the resonant frequency of the circuit under test, some of the dip meter's energy will be absorbed in the coil being tested. The meter needle indicates the strength of the dip meter oscillator output. As some of the energy is absorbed into the

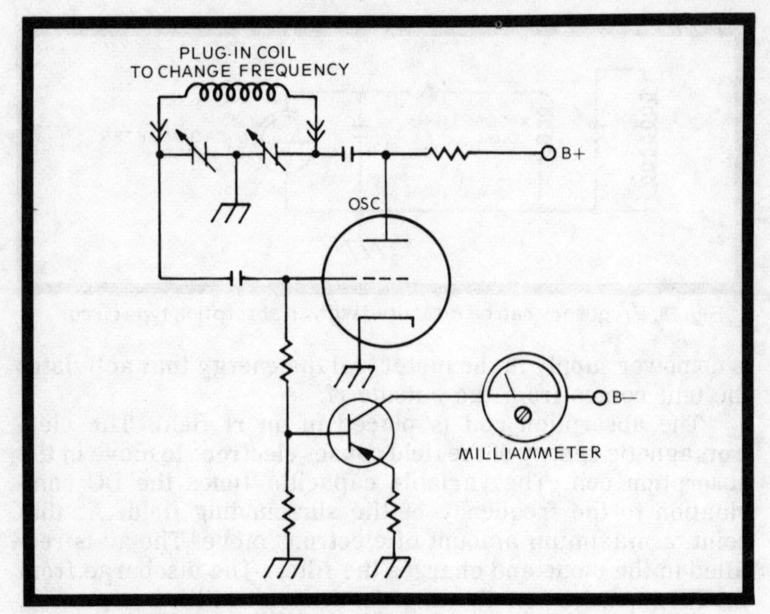

Fig. 99. A dip meter applies energy to the coil under test. The degree of meter dip indicates the amount of energy absorbed.

device under test, the oscillator energy drops and the meter "dips."

The dip meter also will indicate a loss of energy at frequencies in the harmonic range of the device under test. Therefore, when making the dip-meter calculations, couple the meter coil to the device as loosely as possible, just close enough to give a perceptible dip to the needle. If it is in too close, it is possible to get an incorrect indication because the harmonics are stronger with close coupling.

62 CAPACITOR TESTER (DIP METER)

A dip meter can be used to measure capacitance. The only other accessory needed is a coil with a known inductance (Fig. 100). A standard value is 5 uH. Attach the capacitor in question to the coil and form a tuned circuit. Then bring the dip meter in range and let the tuned circuit absorb some of the radiated energy. Tune the dip meter till the maximum dip occurs. The tuned circuit at that point is absorbing a maximum of energy from the radiation. The frequency on the dip meter control is noted. The inductance is known. The capacitance can be calculated, but it is easier to refer to a chart that lists capacitance and inductance values for various resonant frequencies.

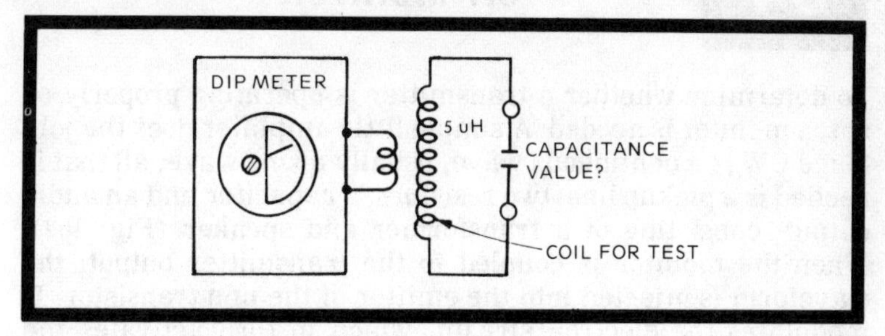

Fig. 100. With a dip meter and a known inductance you can measure the value of unknown.

INDUCTANCE TESTER (DIP METER) 63

Just as the dip meter can be used to measure capacitance, it is useful in determining inductance. But instead of using an inductance as a standard, an ordinary 5 percent 100 pF capacitor is used. The coil in question is attached to the capacitor to form a tank circuit (Fig. 101). Then the dip meter is brought close and the best dip produced. The frequency and the capacitance are known. The inductance can be calculated from a handy chart.

If you want to check the accuracy of the dip meter, take the 5 uH coil and attach it to the 100 pF capacitor. According to calculations the tuned circuit formed should resonate at 7100 kHz. Just be careful to keep the capacitor leads as short as possible so they do not introduce unwanted inductance.

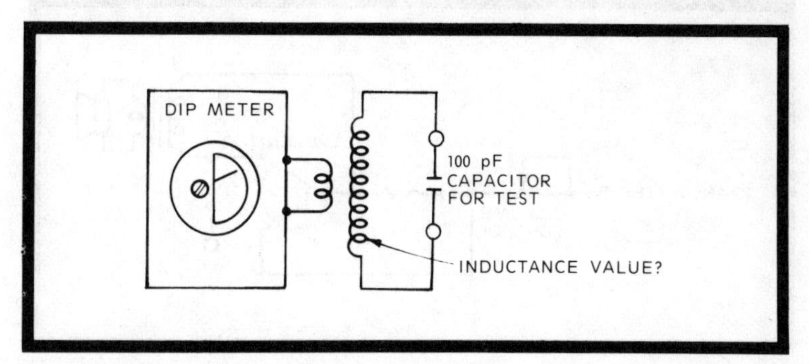

Fig. 101. With a dip meter and a known capacitance you can measure an unknown inductance.

64 CW MONITOR

To determine whether a transmitter is operating properly or not, a monitor is needed. A simple little amplifier does the job. Since CW is a continuous wave, usually a sine wave, all that is needed is a pickup line, two resistors, a capacitor and an audio output, consisting of a transformer and speaker (Fig. 102). When the monitor is coupled to the transmitter output, the waveform is injected into the emitter of the npn transistor. It modulates the electron stream, which in turn activates the audio transformer.

The transformer and speaker form a resonant circuit which dampens the waveform. The result simply makes the voice coil in the speaker emit a gutteral sound. This audio is not modulation on the CW, but simply a result of the dampened CW. The noise occurs during the presence of CW and goes off if there isn't any CW. This is enough to provide the indication that the CW is being generated, which is all the monitor is supposed to do.

The emitter resistor is the volume control, since it determines how much CW is allowed into the transistor. The other two components, the base resistor and the coupling capacitor, are critical as to their values. If they are incorrect values, they could change the CW so that it is of a waveform that is not dampened out by the resonance of the speaker and output transformer. The way to determine the correct values is by trial and error, since the resonance cannot be calculated easily with different transformers and speakers.

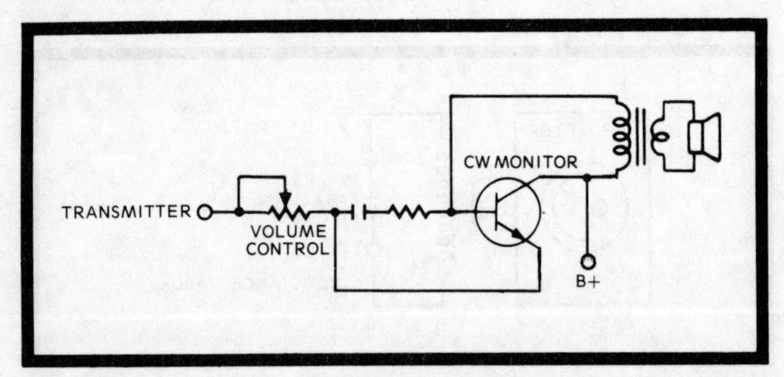

Fig. 102. This CW monitor circuit produces an output due to the resonance between the output transformer and the speaker.

CARBON MICROPHONE CIRCUIT 65

The carbon microphone has been an old standby for many years. It can be attached to the base of an npn transistor through a series filter (Fig. 103). The emitter resistor is bypassed with a large filter, since audio frequencies will not be bypassed unless a large capacitance is used. B-plus is applied to the collector and also to the microphone. A volume control regulates the amount of B-plus that can get to the microphone.

The carbon microphone is a resistive device. The carbon is packed loosely in granular form against a metal diaphragm. As you speak into the mouthpiece, the air waves move the diaphragm, which changes the packing of the carbon granules. This varies the resistance in unison with the speech.

Electrons come from ground into the varying resistance, then pass through the series RC filter, the volume control, and on into B-plus. The audio variations change the positive charge of the series filter, which is coupled into the base. The variation in the base charge modulates the emitter-to-collector electron flow. The modulation is developed across the collector load resistor and is coupled capacitively to the next amplifier stage.

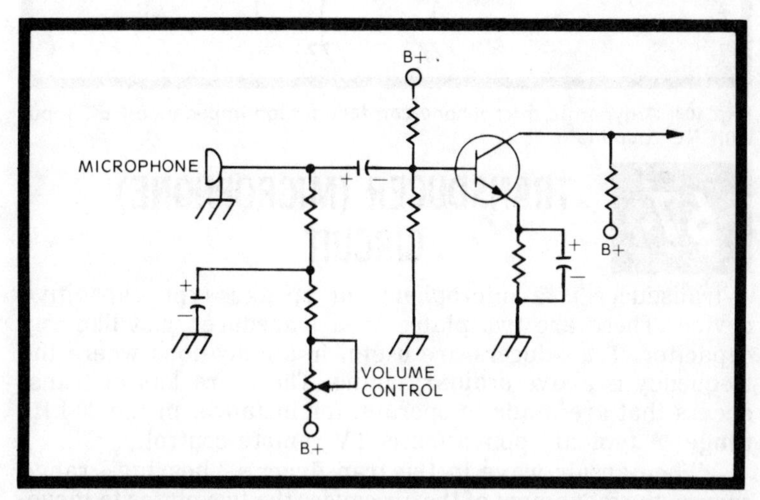

Fig. 103. The output of a common carbon microphone can be fed into a transistor with an RC coupling setup. Notice that B-plus also is applied to the microphone.

66 DYNAMIC MICROPHONE CIRCUIT

A dynamic microphone is nothing more than a speaker being used as a microphone. The voice coil is moved by the speaker diaphragm through a magnetic field. Electrons in the coil move in accordance to the audio as the wire passes in the field. If an FET is used as the microphone amplifier (Fig. 104), the audio is coupled through a capacitor to the gate. The source draws electrons from ground, filters the current flow with an RC bypass, and sends it into the channel, where it is modulated by the varying gate voltage. Modulated dc emerges at the drain. The signal is developed across the drain load resistor and is coupled capacitively to the following amplifiers.

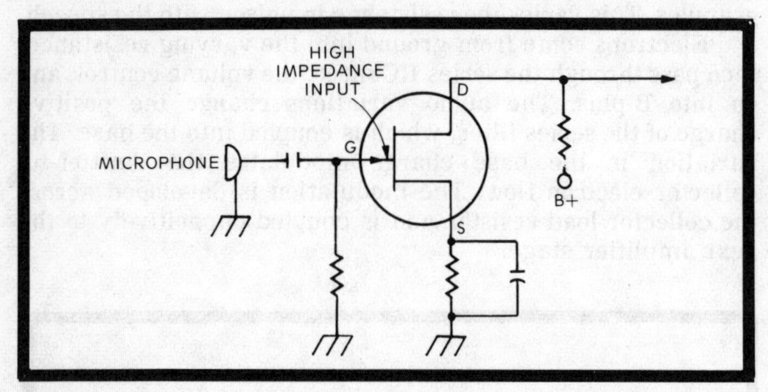

Fig. 104. A dynamic microphone can feed a high-impedance FET input with RC coupling.

67 TRANSDUCER (MICROPHONE) CIRCUIT

A transducer is a microphone- or speaker-type capacitive device. There are two plates on a transducer just like any capacitor. Transducers are useful in applications where the frequency is above ordinary audio. There are lots of transducers that are made to operate, for instance, in the 30 kHz range. A typical application is TV remote control.

When an air wave in the transducer's "hearing" range occurs, the movement of the air causes the two plates to move. As they move, the distance change between the plates varies the capacitance. A dc voltage is applied to the transducer plates (Fig. 105). As the air waves move them, the dc level varies according to the signal, which is typically a sine wave without modulation. The varying dc actuates the control circuits.

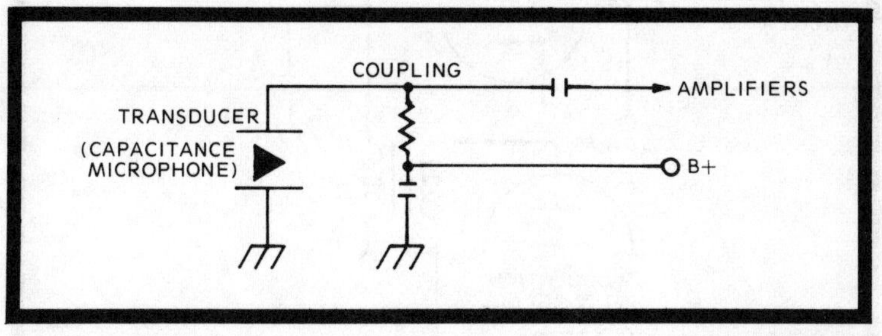

Fig. 105. A capacitor microphone transducer is utilized chiefly in ultrasonic applications.

PLATE MODULATION 68

The most common modulation system is plate modulation of a Class C triode amplifier tube (Fig. 106). The triode grid is biased well below cutoff. An rf sine wave is applied to the grid. Because of the high negative grid bias, the tube does not conduct during most of the sine wave. Only about 100 degrees of the most positive part of the sine wave turns on the electron stream. As a result, the plate current consists of short bursts of rf energy. This produces a high-efficiency output that is fed to an antenna coupling transformer. Due to the action of the transformer, the plate current bursts are made into the output sine wave.

Meanwhile, intelligence, typically audio, is developed in a pair of triodes in push-pull. The output transformer secondary is connected to B-plus and a center tap of the rf amplifier plate transformer primary. The audio adds to the rf and produces sidebands. This is plate modulation.

Modulation Envelope

There are three common ways that intelligence can be impressed on an rf carrier wave. They are amplitude modulation (AM), frequency modulation (FM) and phase modulation (PM).

Amplitude modulation occurs in the circuit just discussed. If it is audio that is doing the modulating, the sounds can be reproduced well in a band of zero to 5000 Hz. If the carrier

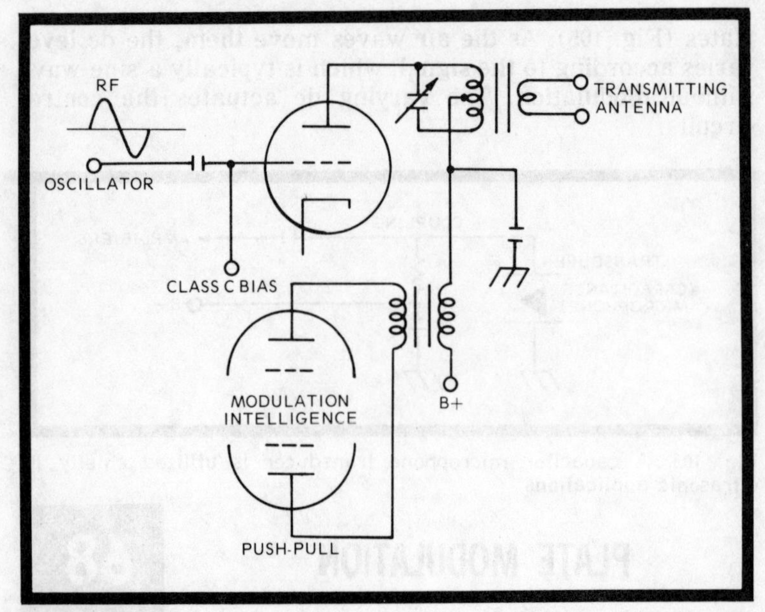

Fig. 106. Plate modulation is the most common method used in tube-type transmitters.

frequency is 610 on the dial (610 kHz), then the carrier will be moved 5 kHz above and below 610. That is a high of 615 and a low of 605, the normal AM broadcast modulation range.

When producing the carrier plus the sidebands, it is desirable to fill up as much of the modulation envelope as possible. This is spoken of as percentage of modulation. A certain amount of valuable power has been expended to produce an envelope with high amplitude. When the carrier is modulated, if the modulation is only 50 percent of the rf amplitude, the unused part of the rf cycle represents wasted energy. Therefore, it is useful to set up the modulator circuitry so that the entire rf envelope is modulated, up to 100 percent.

If modulation exceeds 100 percent, then overmodulation takes place. This can cause distortion of the sounds that you want to reproduce in the receiver; 100 percent is the desired percentage of modulation.

69 SCREEN MODULATION

It is possible to use the other elements in a tube to inject the intelligence. The cathode, control grid, screen grid, and suppressor grid can all introduce moduation to the rf carrier wave being amplified. Screen grid modulation is the most preferred of these and all operate in a similar manner.

In a pentode tube, the rf enters through the control grid and turns on the tube. The push-pull stage producing the audio drives the screen grid through a coupling transformer (Fig. 107). As the electrons in the pentode pass the screen, the audio modulates the intensity of the electron stream. The modulated rf is developed across the load in the plate, which is the coupling transformer.

In screen modulators a small rf bypass capacitor gets rid of any stray rf that might be coupled to the screen through interelectrode capacitance. The bypass is a very small capacitance that has a low resistance to the rf but almost infinite resistance to the audio passing through the screen grid.

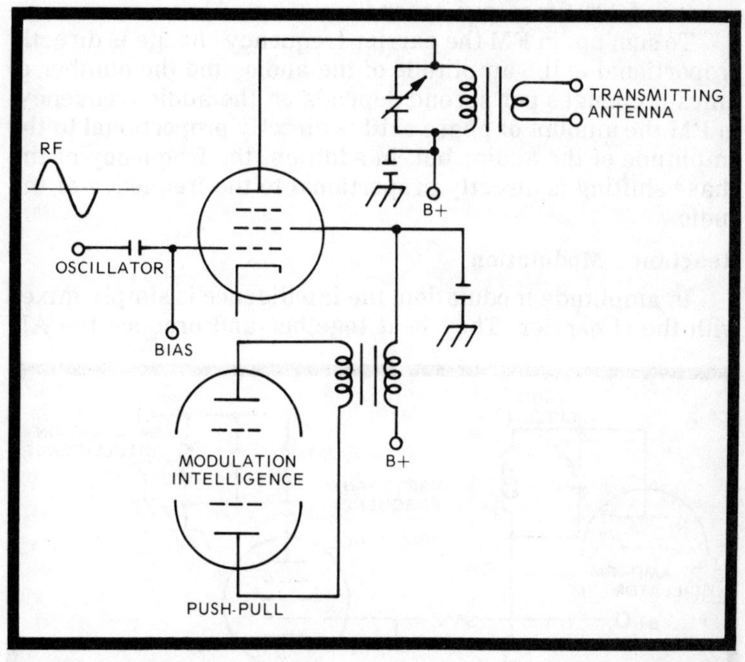

Fig. 107. Screen modulation also is used. Intelligence can be applied to the control grid or cathode.

FM AND PM MODULATION CIRCUIT 70

There is considerable confusion between FM and PM. That is because they are not two different procedures, but variations on the same procedure. The same modulator circuit, called a reactance modulator, produces either one (Fig. 108).

In FM, the amplitude of the audio is made to vary the frequency of the rf carrier. When the audio signal is positive it makes the rf higher in frequency. When the audio signal is negative it makes the rf lower in frequency. The amplitude of the carrier stays constant. Only the frequency deviates.

In PM, the amplitude of the audio is made to vary the phase of the rf carrier. When the audio is positive it shifts the phase forward from the unmodulated position. The amount of forward shift is determined by the audio amplitude. When the audio is negative it shifts the phase backward from its unmodulated position. The amount of rearward shift is determined by how far negative the audio sine wave is. The frequency of the audio determines how fast the phase shifting takes place. For a 400 Hz note, the phase shifts forward and rearward 400 times per second.

To sum up, in FM the carrier frequency change is directly proportional to the amplitude of the audio, and the number of times it changes per second depends on the audio frequency. In PM the amount of phase shift is directly proportional to the amplitude of the audio; but, in addition, the frequency of the phase shifting is directly proportional to the frequency of the audio.

Reactance Modulation

In amplitude modulation, the intelligence is simply mixed with the rf carrier. They beat together and produce the AM

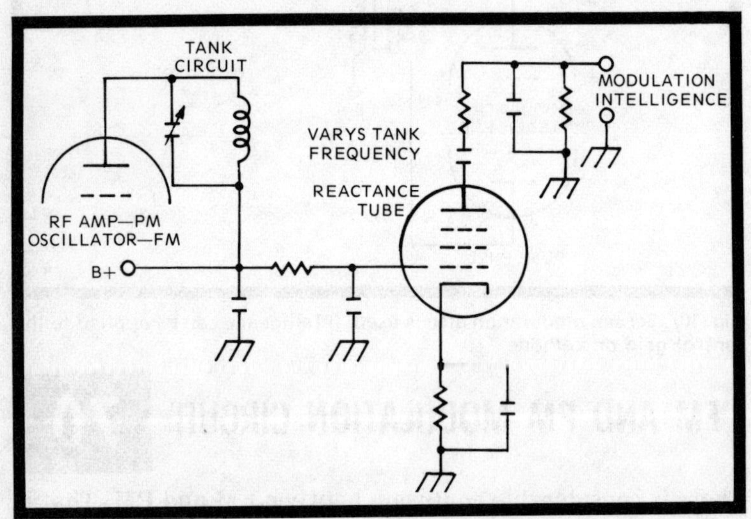

Fig. 108. A reactance tube is used to frequency modulate a carrier.

envelope. In FM or PM, it's not that simple. The intelligence must be added to the rf to change its frequency or phase in accordance to the desired modulation. This can be done by using a variable inductance or capacitance in a mixer-type amplifier. This is called reactance modulation.

The idea is to apply the intelligence to the reactance tube. Its output then becomes a variable inductance or capacitance according to the proportions of the intelligence. The reactance tube functions as the capacitor or inductance in a tuned circuit. The changes then change the conduction of the modulator tube. For FM, the reactance circuit is part of the oscillator tank circuit. The oscillator frequency varies, producing the FM output. For PM, the reactance circuit is part of the amplifier tank. That way the oscillator still stays on frequency and the phase shifts do not take place until the amplifier is detuned from its resonant frequency.

FM and PM, by their very nature, occupy much larger areas of the frequency spectrum than does AM. The carrier waves are usually well in the megahertz range, while AM is typically much lower. Since the megahertz range is used, a frequency response up to 20 kHz can easily be transmitted and received. Since most interference is amplitude modulated and good FM or PM receivers are insensitive to AM, the strong FM transmission is relatively noise free.

The detectors used in FM and PM receivers are much more complex than their AM counterparts. They require many more components and extensive delicate alignment. They are called discriminators, ratio detectors, and phase detectors. In comparison, an AM detector is simply a diode and RC network.

DISCRIMINATOR CIRCUIT 71

While an AM detector must be sensitive to variations in amplitude, an FM detector must be able to convert variations in frequency to the original modulation.

The typical FM discriminator uses two diodes in parallel (Fig. 109). The cathodes are tied together through a pair of capacitors and resistors in series. The anodes are attached to the ends of a coupling transformer secondary. The center tap in the secondary is capacitively coupled to the primary and to the center of the RC pairs. The coupling transformer is tuned to the center frequency of the incoming FM i-f signal. The two resistors are the cathode loads, while the two capacitors have a value that bypasses the i-f signal to ground. They actually

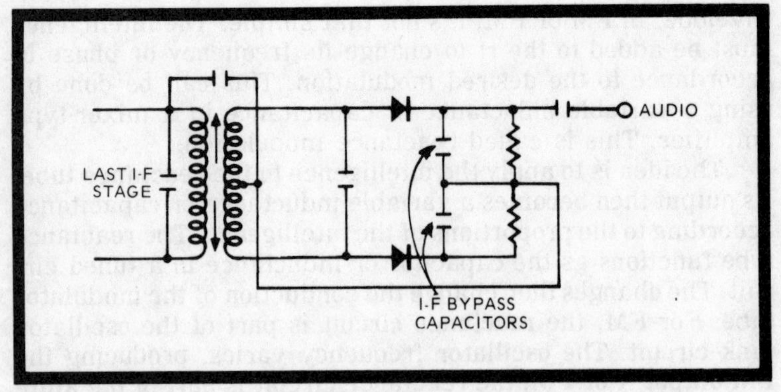

Fig. 109. Frequency modulation signals are typically detected in a discriminator circuit.

get rid of the i-f in the same manner the AM detector bypass gets rid of the i-f. (The two capacitors are the only components that have an equal counterpart in an AM detector.)

No B-plus voltage is applied to the diodes. The only way electrons are able to flow is when they are repelled or attracted by voltage variations caused by the incoming i-f signal. When the i-f is developed across the primary at the center frequency (no modulation), it causes equal and opposite of amounts of electrons to move back and forth in the secondary. Simultaneously, the i-f is coupled to the center tap of the secondary through the coupling capacitor. Also, some of the i-f is coupled to the center of the two load resistors. Voltages are developed on the two anodes. The top anode voltage is equal to the sum of the primary and top half of the secondary. The bottom anode voltage is equal to the sum of the primary and the bottom half of the secondary. Since the i-f is at center frequency, these two anode voltages are equal. Equal amounts of current are drawn from the load center tap through each load resistor and through the diode.

The total voltage across the full load is equal to the difference between the voltages across the load resistors and will have the polarity of the greater of the two voltages. Since the two voltages are equal, the audio output voltage is zero at center frequency. As the i-f rises above center frequency, or deviates in a positive direction, the voltage on the anode of the bottom diode becomes larger than the voltage on the top diode. Therefore, there is a greater voltage drop across its load resistor. As a result, the output voltage goes negative. As the i-f goes below center frequency, or deviates in a negative direction, the voltage on the anode of the top diode becomes larger than the voltage on the bottom diode. Consequently, the voltage drop across its load resistor is higher and the output voltage goes positive.

The output of the discriminator is the audio that frequency modulates the rf carrier. Actually, this discriminator circuit is used in lots of applications other than audio detection. Typical are afc and horizontal phase detectors.

PREEMPHASIS & DEEMPHASIS CIRCUIT 72

In FM broadcasting, the higher audio frequencies between 1,000 and 15,000 Hz are preemphasized at the transmitter. This is because noise is usually in that range. If the higher frequencies are made stronger, the noise is rejected. As the 15 kHz range is reached, the signal can be as much as the 16 dB stronger than the lower frequencies. This means the transmitted FM signal deviation from center frequency is much greater at the higher frequencies. This is called preemphasis.

Preemphasis is achieved before modulation by an RC circuit. The time constant of the RC network is 75 microseconds.

In the receiver a deemphasis RC network is needed to restore high-frequency audio to the proper level. Typically, the RC is a 75K resistor in series with a .001 uF capacitor to ground (T = 75K x .001 = 75 microseconds). If the deemphasis network should fail or be removed from an FM receiver, the highs will become overpowering and the radio will screech.

As the signal comes out of the discriminator it passes through the RC network. The low frequencies pass normally since the capacitor reactance is high. As the frequency increases, the capacitor reactance decreases and bypasses the highs to ground. The higher the frequency the more of the signal is bypassed to ground. The values are critical. Exact or close values should be used as replacements.

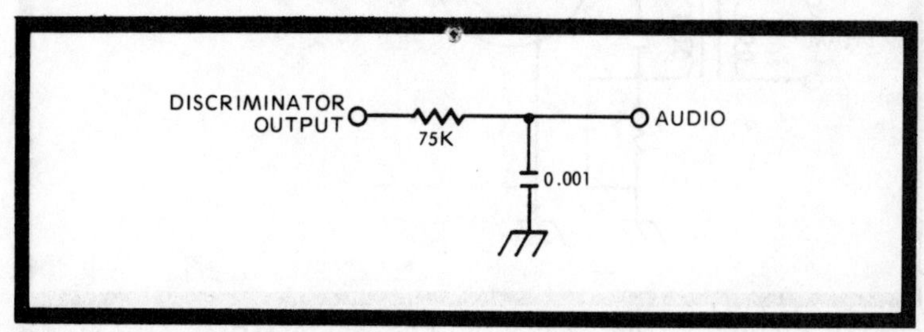

Fig. 110. A deemphasis network is used after an FM detector to restore normal frequency response.

73 LIMITER AMPLIFIER

The main job of a limiter amplifier is to maintain a steady input to a discriminator. A discriminator is not only sensitive to FM but will also detect amplitude variations. This is undesirable because the only AM in an FM broadcast is noise. The limiter circuit clips the tops and bottoms off the carrier wave, leaving only the FM variations (Fig. 111). The limiter is usually the last i-f stage. It provides some amplification but not too much.

A limiter can be a tube, transistor, or an FET. In a tube the B-plus is applied in very small amounts. This restricts the available operating area on the linear part of the characteristic curve. The tube is biased at Class A, or the center of the linear curve. Since the limiter is the last i-f, the signal is quite strong. As the signal swings in a positive direction, it causes the gred to draw a current at the peak of the signal, thus signal peaks are clipped and do not appear in the plate circuit. As the signal swings in a negative direction, the tube goes into cutoff about half way toward the negative peak and the tube stops conducting. The bottom of the signal thus does not appear in the plate.

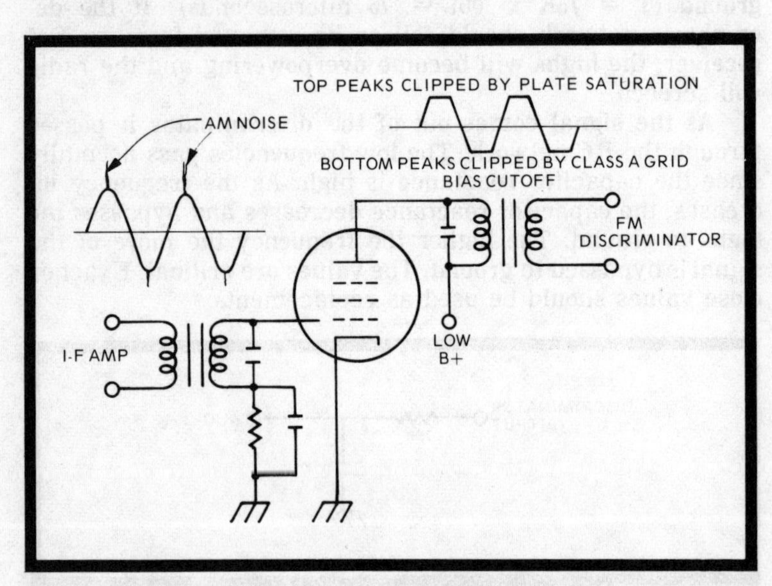

Fig. 111. A limiter amplifier follows the i-f stages in an FM receiver to get rid of noise peaks.

In the preceding i-f stages the entire carrier including the noise pulse is passed and amplified. In the limiter, the peaks of the carrier are clipped off and only the carrier frequency variations pass on to the discriminator. Some receivers have two limiters to further insure clipping.

RATIO DETECTOR AMPLIFIER 74

The ratio detector circuit is quite like the discriminator and does the same job. Its main distinction is that it does not need a limiter stage before it, since it is not sensitive to AM.

Looking at the circuit in Fig. 112 the first notable difference is that one of the diodes has been reversed (the bottom one). The anode is connected to the top cathode. Next, we notice that the two load resistors are replaced by a single load resistor across the two bypass capacitors, which still get rid of the carrier wave. Lastly, there is a large filter capacitor across the load resistor. The load resistor and filter capacitor have a large time constant between a quarter and a half a second. Any amplitude variations in the carrier are completely filtered out in the capacitor's charge.

Since there is only one load resistor across the two capacitors and since the load voltage is equal to the sum of the two voltages in the capacitors, the load resistor does not develop the audio. The load resistor maintains a constant voltage that is dependent only on the carrier strength. The load resistor voltage changes only when different FM stations are tuned in. The audio voltage appears at the center tap of the capacitors. This voltage depends only on the phase of the voltage in the ratio detector transformer and not on the amp-

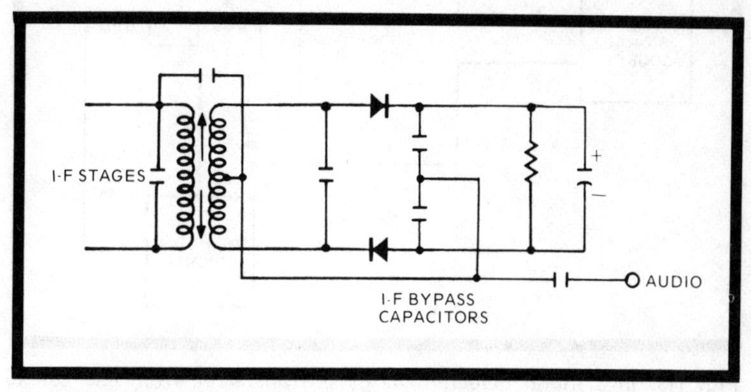

Fig. 112. A ratio detector circuit is similar to a discriminator.

litude of the signal. The two capacitors charge, filter the carrier, and the center voltage varies according to the frequency deviation. Then the signal is coupled to the de-emphasis network through a blocking capacitor.

75 METAL DETECTOR CIRCUIT

The detector-type circuit is widely used in metal detector devices. It works like this. An ac signal is applied to a pair of coils in series (Fig. 113). Around the two coils a magnetic field is developed. A second pair of coils is placed near the first pair, but at right angles to the first pair. In this position little or no mutual coupling can take place between the first pair. As long as the two coils remain untouched, no energy is coupled from the primary to the secondary.

The primary can be on one side of a doorway, while the secondary is on the other side of the doorway. Should a person walk through this doorway with a metal gun in his pocket, the gun becomes an iron core to the two windings and permits some energy to be coupled from primary to secondary.

The circuit is so arranged that the output of the secondary controls the bias on a tube or transistor. As more energy is coupled into the secondary, the balance of the mutual inductance bridge is upset and the bias is changed. The tube conducts, causes a relay to close and sound an alarm. The presence of a gun has been detected!

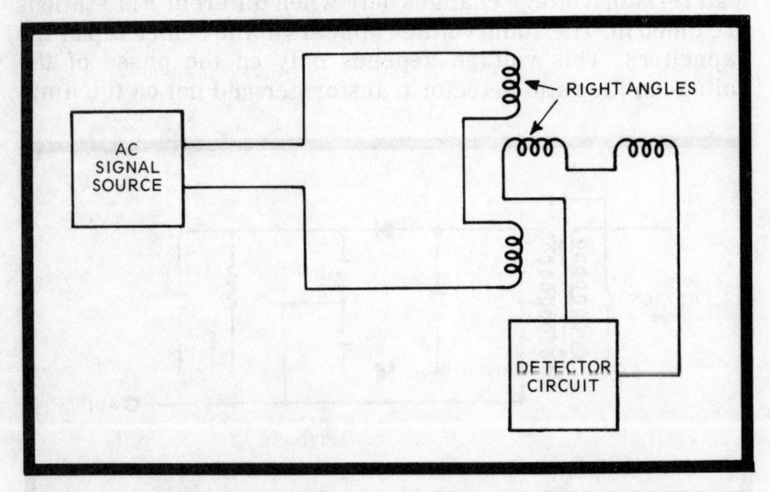

Fig. 113. In a metal detector energy is transferred from one coil to another when the metal forms a type of core.

Bridge

A mutual inductance bridge is a balanced circuit (Fig. 114). The input is the ac from the primary. This develops electron movement in the secondary. In the case of the metal detector, this is a very small amount of current. For all practical purposes it is near zero. A capacitor and three resistors are placed in series with the secondary and the circuit is completed back to the secondary.

The small amount of ac is developed across the components. The trigger circuit input is connected to the junction between the resistor and capacitor and the resistor and coil. The amount of ac and dc in the trigger input is designed to maintain the tube or transistor at cutoff. The circuit maintains a balanced tapped output. When a gun or other metal object enters the airspace between the primary and secondary, the additional energy changes the amount of signal developed across the components and turns the trigger amplifier on.

Sensitivity

The ratios between the resistor values, rather than the individual resistance values, determines how much voltage balance there will be. The amount of resistance determines the size of the metal object needed to upset the bridge balance. If you increase or decrease the amount of resistance in the bridge, you raise or lower the sensitivity of the bridge.

Sometimes it is desirable only to have the alarm go off when large pieces of iron disturb the field. Other times it is desirable to have the alarm go off no matter how small the piece of iron is. Adjustment of these resistances represent a sensitivity control.

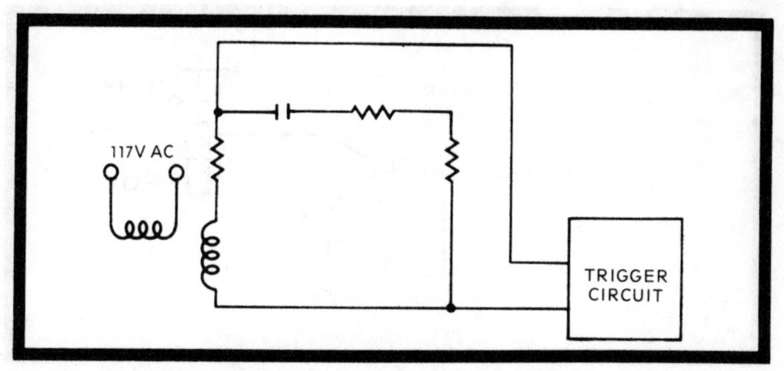

Fig. 114. A bridge circuit can be used in the secondary of a metal detector. The metal detected upsets its balance.

Frequency

The frequency used in metal detectors varies all the way up and all the way down the spectrum. Some detectors use ordinary 60 Hz ac. They can then use a full-wave circuit to produce a frequency of 120 Hz. Other metal detectors, especially portable types, have self-contained oscillators. Mostly low audio frequencies are used, but don't be surprised to find detectors with oscillators well up into the rf range.

The alarm can be silent, with either a light or a meter in place of an audible device. If a meter is used, it is simply attached to the bridge and gives a voltage reading according to the size of the metal that causes the mutual coupling.

76 TRIGGER CIRCUIT

An npn transistor with an iron-core coil as the collector load will function as a trigger circuit. The input can be resistive. In the circuit in Fig. 115, electrons do not flow normally since the input has a reverse bias. As a signal is developed in the preceding circuit, a voltage arrives at the input and overcomes the reverse bias. The transistor turns on. The collector current through the coil creates a magnetic field. Near the iron core is a relay armature which is attracted to the core. The relay movement closes a switch and the alarm goes off. Usually, the switch remains closed until the alarm is turned off by hand.

The base resistor can be made adjustable so that the trigger can be set to go off at different levels for different applications.

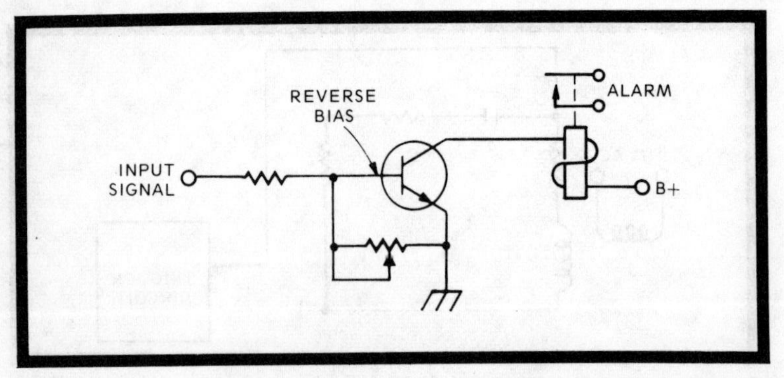

Fig. 115. A reverse biased npn can be used as a trigger circuit. It conducts when a signal causes forward bias.

PROXIMITY CIRCUIT 77

A proximity detector is unlike the metal detector in that it can detect any object, not just metal. It starts an alarm when an object, such as a burglar, enters the area it is designed to protect. While a metal detector depends on a metal object to form an iron core between coils, a proximity detector operates when an object changes the capacitance of a capacitor.

In operation (Fig. 116), a certain amount of capacitance is developed between two plates. If the ground is one plate and a big piece of metal such as wire is the other plate, a certain value of capacitance is established. Should an object come between the plates, the air dielectric is disrupted by the object and the capacitance changes. The change of capacitance is detected in a bridge and an alarm is triggered.

An ordinary tuned-plate tuned-grid oscillator is used in the circuit. In the plate tank a tap is taken off. This is the sensor

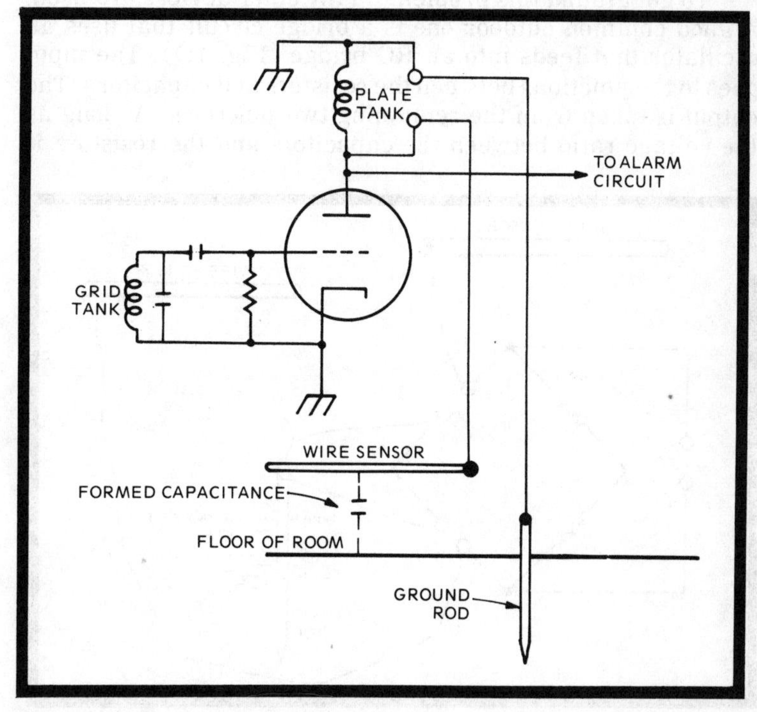

Fig. 116. A proximity detector senses the presence of an intruder by a change in capacitance.

connection. While the oscillator is running, an ac current develops between the sensor and ground. The air between the sensor and ground acts as the dielectric of a capacitor, with a definite amount of capacitance. The amount of capacitance is determined before hand to produce the oscillation.

As an intruder walks into the invisible, weak field, he changes the capacitance by interfering with the air dielectric. This detunes the tank and causes the oscillator to stop running. This load on the oscillator causes heavy current to flow. That changes the bias on a trigger circuit, which turns on an alarm.

78 PROXIMITY BRIDGE CIRCUIT

The preceding circuit is inexpensive and is useful indoors, especially in a constant temperature-humidity atmosphere. However, for outdoor surveillance it is useless. Changes in temperature and humidity can cause as much as a 25 percent change in capacitance. This would trigger the alarm falsely.

To get around this problem, many other devices are used. A good common outdoor one is a bridge circuit that uses an oscillator that feeds into an RC bridge (Fig. 117). The input goes to the junctions between the resistors and capacitors. The output is taken from the remaining two junctions. As long as the voltage ratio between the capacitors and the resistors is

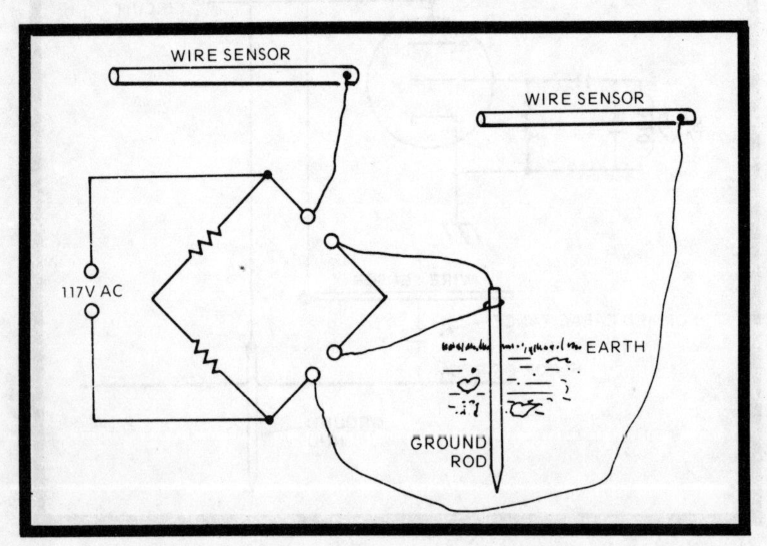

Fig. 117. A bridge circuit proximity detector must be used outdoors. Otherwise, changes in weather will trigger an alarm.

equal, the bridge is balanced and there is no voltage at the output.

One of the capacitors is the sensor and ground. If an intruder walks into the sensing area the capacitance is changed. The bridge loses its balance and sends an ac voltage to the trigger circuitry. In actual use both of the capacitors are sensing wires and ground. Therefore, any temperature and humidity changes will change both capacitances in exactly the same way. Notice the voltage ratio between the capacitance values, not the amount of each capacitance, maintains the balance. Should an intruder walk into either sensor, he changes that particular capacitance, which unbalances the bridge and triggers the alarm.

FET PROXIMITY CIRCUIT 79

A tube or bipolar transistor alone cannot simply measure the voltage across a capacitor. The input impedance in each is so low that it drains off the voltage charge as the measurement is made. An FET, however, has such a high input impedance that it can take a voltage measurement without disturbing the charge across a capacitor. This enables us to construct a proximity device that simply measures the voltage across a capacitor. An FET connected across a capacitance sensor will sense a capacitance change due to the voltage change. If the capacitance decreases, the voltage goes up. Should the capacitance increase, the voltage will drop.

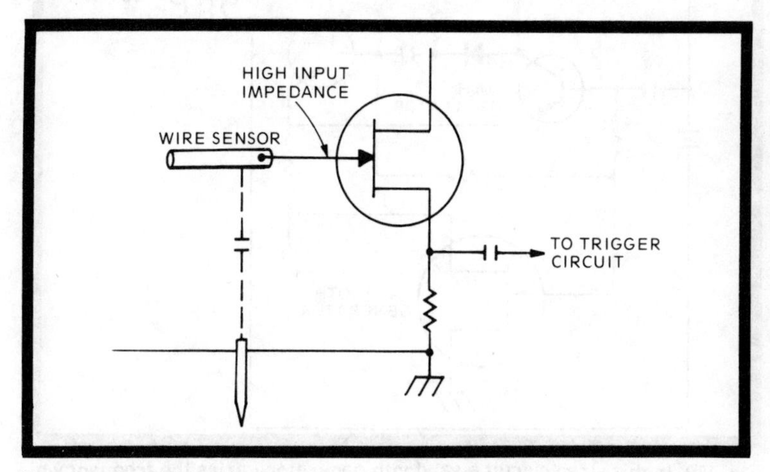

Fig. 118. An FET makes an excellent proximity alarm circuit due to its naturally high input impedance.

A simple FET circuit with the sensor wire attached to the gate is shown in Fig. 118. As the voltage across the sensor wire and ground changes, the change becomes a bias change on the FET gate. The change in electron flow, developed across a drain resistor, is coupled into an amplifier and then into a trigger circuit, which sounds the alarm.

80 SIREN CIRCUIT

One of the most effective type alarms is a wailing siren. Sirens are mechanical devices, but the sound can be simulated with an electronic circuit. An audio oscillator consisting of RC components produces the basic sine wave. The circuit in Fig. 119 uses a pair of tubes or transistors that produce a frequency of about 400 Hz when a switch is closed. The 400 Hz signal is then sent to an audio amplifier circuit that can deliver any desired power output.

Of course, a straight 400 Hz note does not sound like a siren. So a second circuit, consisting of a single tube or transistor, modifies the frequency of the audio oscillator. A sawtooth generator circuit has a voltage output that starts at zero and rises to a designed peak. After it peaks, the voltage amplitude drops off to zero. This change in voltage applied to the bases of the oscillator changes the frequency accordingly.

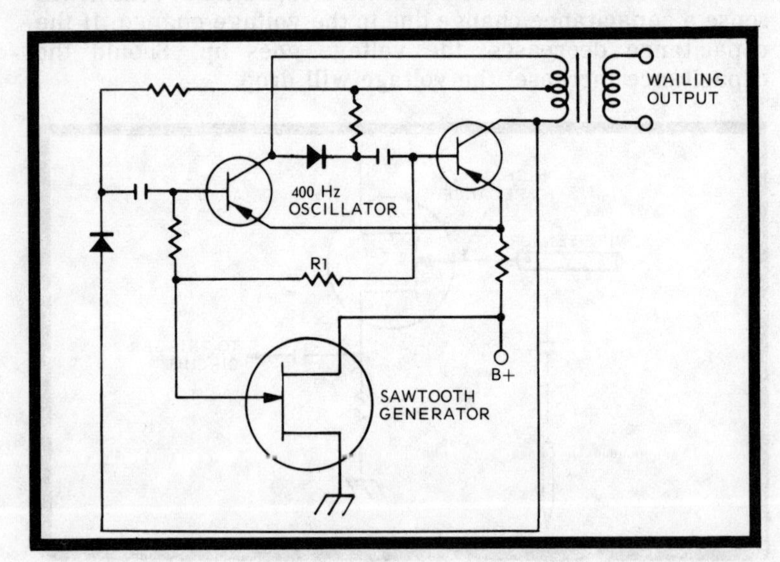

Fig. 119. In this siren circuit a sawtooth generator varies the frequency of an audio oscillator.

As the sawtooth voltage input rises in amplitude, the oscillator frequency increases slightly. As the input reaches its peak, the oscillator hits its highest frequency. Then the sawtooth amplitude drops off to 0V and the frequency of the oscillator drops. This produces a watt or two of wailing audio output from the oscillator. It's then coupled into audio amplifiers for the desired power output.

FM BUG 81

Bugging rooms seems to have become a national pastime. Bugs are hidden in martini olives, coat buttons, etc. A typical bug is a tiny FM transmitter operating in the 50 to 150 MHz range. A good little bug can transmit to an FM receiver as far as 500 feet. It has no revealing wires and can be super miniaturized.

The circuit in Fig. 120 consists of an oscillator that runs at the desired frequency. The oscillator feeds into a 2- or 3-inch antenna. The antenna can be the toothpick in the olive or the pin on the button or corsage. The oscillator consists of a single npn transistor with a tuned collector with feedback to the emitter.

The miniscule microphone is capacity-coupled to a pair of npn transistors that amplify the audio pickup. The second npn is an emitter follower that feeds the audio through a volume control to the oscillator. The audio frequency modulates the oscillator by varying the base bias slightly.

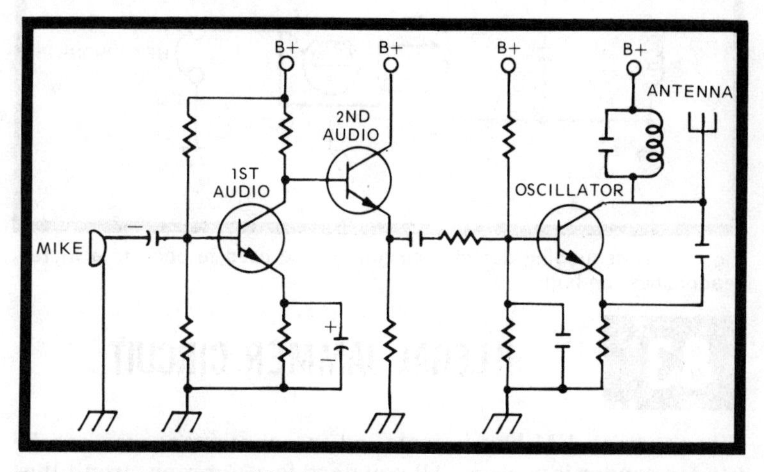

Fig. 120. Short-range FM bugs are easily made in subminiature. This circuit has only three transistors.

82 DEBUGGING CIRCUIT

Since most bugs are FM transmitters operating in the 50 to 150 MHz, a field strength meter that operates on the same frequencies will reveal the presence of such a device. It is helpful if the debugger has a speaker output across the meter so the audio can be heard.

Most field strength meters are sensitivity tuned and the hidden transmitter will be detected only when the FSM is tuned to its frequency. You can work the tiny dial back and forth as you search. However, a broadband FSM that will pick up a wide range of frequencies is better. That way, as you get near the transmitter, you not only hear it in your headphones but you can pinpoint its location as you approach it. The volume keeps increasing and you'll get good howling feedback when you are on top of it.

A simple detector circuit (Fig. 121) with a single diode detector attached to the antenna makes a good broadband debugger. The detector output is injected into an udio amplifier that drives the headphones.

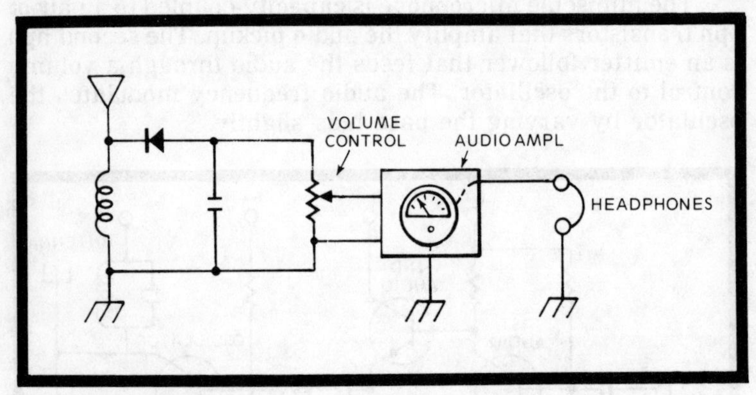

Fig. 121. A debugging detector circuit can be used to operate a meter, headphones, or both.

83 ILLEGAL JAMMER CIRCUIT

Whenever an FM bug is installed, a countermeasure can be used to render it useless. All you need is a jammer circuit that can be hidden inside a ballpoint pen case. Of course, it is against the law to radiate signals without permission, but it is done all the time. It is standard practice for board rooms of large corporations to have good jamming devices running continually. Even if the room is bugged, all the listener gets is annoying static that swamps out any and all conversations.

A jammer is nothing more than a buzzer attached to a coil. When the switch is closed, the buzzer starts operating. The quick opening and closing of the buzzer relay develops an oscillating field around the coil. The field radiates over a very broad band of frequencies as noise. The clandestine receiver picks up the noise. The noise has a high amplitude and the signal-to-noise ratio between the FM bug transmission and the jammer's transmission are about the same. The noise swamps out the tiny bug's transmission.

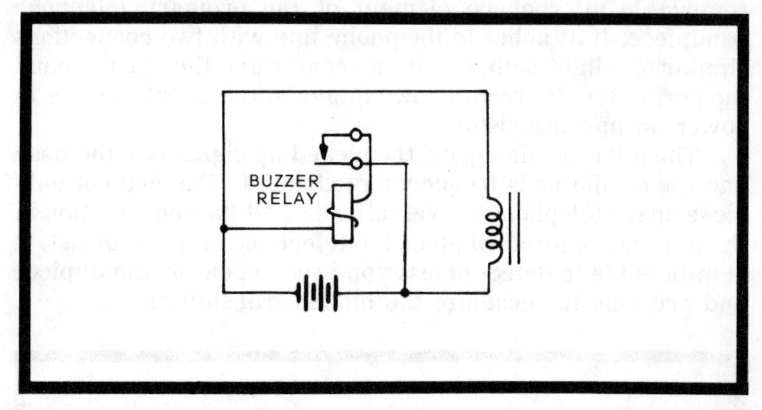

Fig. 122. This illegal jamming circuit is quite simple. A buzzer simply fills the surrounding airwaves with noise.

TELEPHONE TAP CIRCUIT 84

A telephone tap is easily made with nothing more than a high-impedance head set. Simply attach the head set in parallel with the telephone to be tapped. The higher the impedance the less the volume of the telephone conversation will drop. If you want to attach an amplifier and tape recorder, the amplifier should have a high input impedance.

You've noticed how volume drops a bit when your extension telephone is off the hook. This is due to the relatively low impedance of the extension. In order not to be noticeable, a higher impedance type is needed for the secret tap. The tap can be made at the telephone switchboard or on the telephone pole board. The FET, with its extraordinarily high impedance, makes a good high-impedance amplifier.

85 TELEPHONE FM BUG

The telephone tap is convenient when it is installed with the telephone company's permission. A room can be rented and all the elaborate equipment brought in to collect evidence. This is not always possible, though, and private citizens resort to less legal methods. One such way to bug a telephone is with an FM transmitter installed right into the mouthpiece (Fig. 123).

FM bugs come in a case that looks identical to the removable microphone element of the ordinary telephone handpiece. It attaches to the phone line with two connections similar to a light bulb, one in the center and the other around the perimeter. It uses the low voltage of the telephone line to power an npn transistor.

The microphone injects the picked up signal into the base and the oscillator is frequency modulated. This bug not only picks up the telephone conversation but all the conversation in the immediate area, whether the telephone is in use or not. It is impossible to detect unless you break open the mouthpiece and are able to recognize the hidden transmitter.

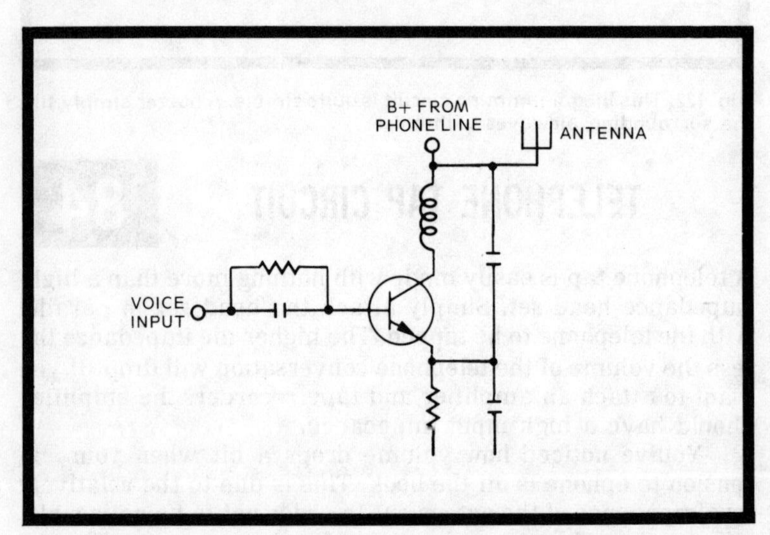

Fig. 123. A telephone FM bug can be obtained in a package identical to the Western Electric mike.

TUBE TUNING INDICATOR 86

The old tube tuning indicator has the familiar round circle with a green piece-of-pie-shaped light. When there is no signal between stations, the pie wedge is maximum, which is about 100 degrees of the circle. As a signal is applied, if it is not too strong, the wedge narrows to about 40 degrees. As a good strong local channel is received, the wedge narrows down to a sliver of just a few degrees. By rocking the tuning knob about center, best tuning is indicated when the green light closes to the minimum number of degrees. The tube itself has two main sections. One is a high-gain triode and the other is a cone-shaped target coated with a green phosphor.

The triode is operated as a dc amplifier (Fig. 124). The cathode goes to ground and the grid is attached to agc through a 1M resistor. A .01 uF capacitor bypasses ac from the grid to ground. The capacitor is an open circuit to the incoming varying dc from the agc line.

When the receiver is tuned between stations, the agc is most positive and provides little or no bias on the grid. As a result, a maximum amount of electrons flows through the grid. When the receiver is tuned to a station, the agc is most negative and provides a high bias for the grid. Little or no electrons flow past the grid. The amount of agc voltage determines the degree of tube conduction. The agc voltage, of course, is a function of signal strength.

The target section has another cathode that is attached to the triode cathode. The plate of the triode is attached to the control electrode located between the target cathode and the target. There is a high resistance between the target and the plate of the triode. The control electrode is shaped like a blade.

When the control electrode is less positive than the target, electrons are repelled by the control electrode, casting a wide shadow on the target. When the control electrode is at the same potential as the target, electrons are not repelled and the shadow becomes narrower.

The target is more positive than the control electrode when the agc is positive. Heavy current flows in the triode, dropping the plate voltage and the voltage on the attached control electrode. Agc is positive during no signal. The target potential is the same as that on the control electrode when the agc is negative. No current flows in the triode, making the target and control electrode potential the same.

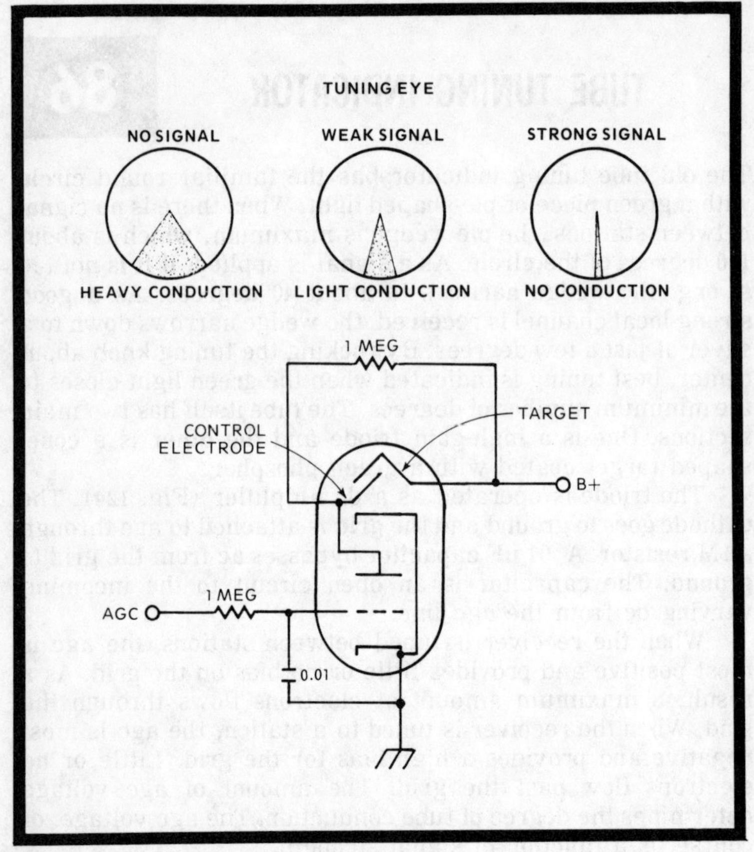

Fig. 124. A tuning-eye tube, with its familiar green pie-shaped light, is a complex triode tube with a target.

87 PHOTO DETECTOR CIRCUIT

In certain types of transistors, if you shine a light on them, electrons will flow. The transistor can be an npn with a collector load and grounded emitter (Fig. 125). The base has no wires connected to it. Instead, there is a lens in front of it that focuses light on it. The light gives the base a forward bias and electrons flow from emitter to collector. The more light, the greater the collector current.

The phototransistor device can be set to measure the normal light in a room, which produces a certain amount of collector current. Should something moving change the light input to the phototransistor, either by adding additional light or cutting off the light, the load current changes accordingly. The current change can be used to trigger circuit an alarm.

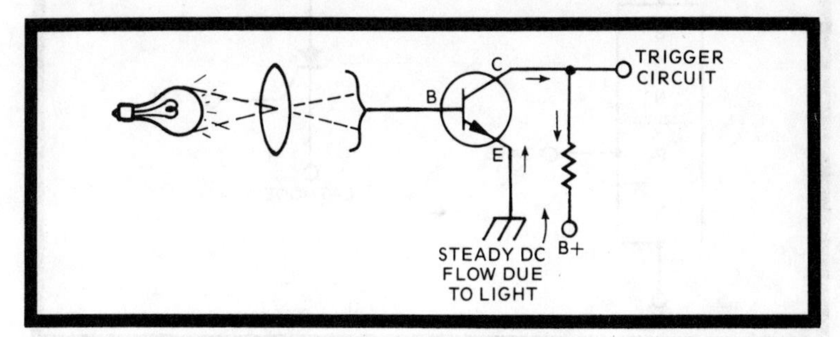

Fig. 125. A photo detector transistor develops forward bias when light is focused on its base.

SCR SWITCH CIRCUIT

88

The term, silicon controlled rectifier (scr), is a misleading name. Rectifiers are among the simplest type of devices, but an scr is rather complex. It is a combination transistor and rectifier. It has two pieces of p material and two pieces of n material welded together, with a cathode at one where the n material is and an anode at the other end where a piece of p material completes the chain of building blocks (Fig. 126). One of the pieces of p material is a gate. If you consider the scr as an npn transistor with another piece of p material attached to the collector piece of n material, you have it. The npn, with its cathode and gate, acts just like an ordinary transistor. The extra piece of p material attached to the n material collector acts as an ordinary solid-state diode.

Electrons enter the cathode and move across the first pn junction. The electron movement can be turned on, turned off, and modulated at the gate; then the electrons cross over the second and third pn junctions. Electrons leave via the anode.

There is no ordinary pnp type scr where electrons can go the other way. These units are scrs are all of the npn plus a p type. They can be used with forward or reverse bias according to the application. Usually, an scr is operated with forward bias. A positive voltage is applied at the anode and a negative voltage at the cathode. In an ordinary diode with one pn junction, that type of bias produces an electron flow from the cathode, across the pn junction and on to the anode. In an scr there are three pn junctions. A low forward bias produces no current flow. That's because the middle junction, between the

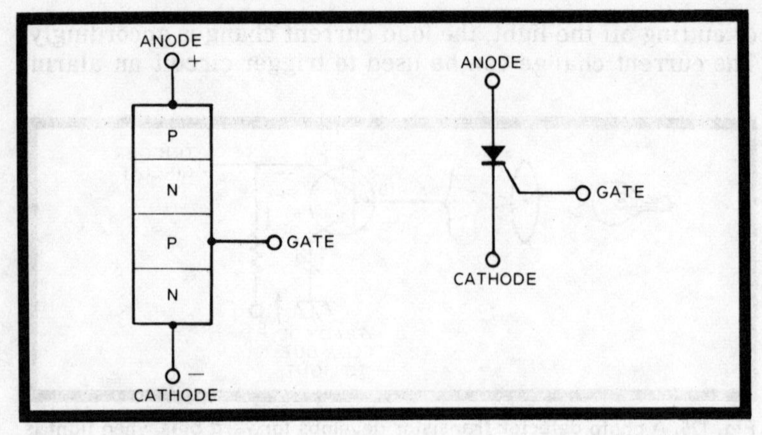

Fig. 126. An scr is an npn transistor with an extra piece of p material attached to the collector piece.

inside p and n materials, is reverse biased. The other two junctions are forwarded biased.

In order to overcome the middle junction bias, the forward bias must be raised to a point where the middle junction reverse bias is increased to its knee voltage (as was described in the zener diode circuit). When the bias is high enough to reverse bias the middle junction far enough, the scr turns on. Electrons flow heavily from cathode to anode. It's as if a zener diode was hidden in the center of the pnpn device. This conduction voltage is called **forward breakover.**

There is another sophistication in the scr. It has a gate. The gate affects that middle junction. By raising the forward bias on the gate, the middle junction knee voltage is moved closer to zero from its high minus value. In other words, when gate current flows it reduces the necessary forward breakover voltage. For a fixed forward bias, the gate determines when current in the scr should flow. The gate current turns the scr on, and no or less gate current turns the scr off. If you apply a sine wave to the gate, the scr will conduct at the peaks of the sine wave. By judicious application of forward bias and gate signals, the scr becomes an excellent electronic switch. In reverse bias, the scr acts like an ordinary diode in reverse bias.

89 SCR LOAD ADJUSTER

An scr can be installed into a circuit to do double duty. One, it can rectify incoming ac, just like an ordinary diode. It does this by conducting during the positive half cycle and cutting off as the negative half cycle creates a reverse bias. Then, the scr can vary the amount of energy that is allowed to go to the load. A rheostat can do the same job by absorbing all the excess energy.

Suppose you want to supply varying amounts of energy to a circuit, heavy sometimes and small other times. All you have to do is attach a rheostat and adjust it for the required amounts of energy. All the energy is delivered to the rheostat and only the desired amount leaves the rheostat.

With an scr there is no wasted power. Energy is delivered to the scr from the source. The scr is forward biased but doesn't conduct until the gate passes current. A gate pulse is injected. The scr conducts only during the duration of the pulse. If the pulse is adjusted so that it arrives simultaneously with the positive ac peak, a maximum amount of current will flow at that time. Should the pulse be adjusted so it arrives during the last few degrees of the positive alternation, a minimum amount of current will flow at that time. If the pulse arrives during the negative swing of the ac, no current will flow. By adjusting the timing and the duration of the gate pulse, large amounts of potential energy can be controlled without large losses introduced by the rheostat.

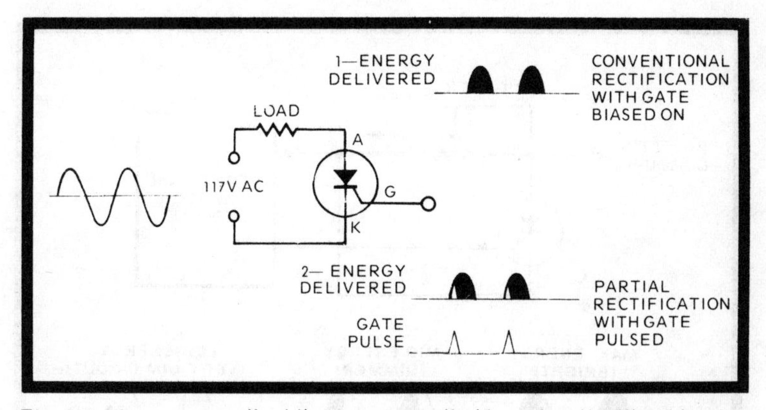

Fig. 127. An scr can adjust the energy applied to a circuit without having to dissipate any energy as a rheostat does.

LIGHT DIMMER CIRCUIT 90

It is a luxury feature in a home to be able to brighten or dim incandescent lights by adjusting a knob. An scr provides an efficient means of accomplishing the task. The scr gate input pulse is adjusted by a potentiometer. The gate input pulse in turn determines what segment of the ac half cycle the scr will conduct. If the scr conducts during a large portion, the light will be bright. Conduction during a smaller part of the half cycle reduces electron flow and dims the light.

An RCA kit circuit operates like this (Fig. 128). The scr is connected to a potentiometer and capacitor, which forms an RC time constant that takes a definite time to charge. If the capacitor charges quickly, the scr will be triggered early in the input half cycle. If the capacitor charges slowly, the scr will be triggered later in the half cycle. If the capacitor does not reach its full charge during the input half cycle, the scr will not be triggered at all. Therefore, at minimum resistance the scr fires early in the half cycle. The maximum amount of energy reaches the incandescent load and the light is bright.

At mid-resistance the scr fires later in the cycle. A smaller amount of energy reaches the load and the light is dimmer. At maximum resistance the scr does not fire and no energy gets to the load. The light is out. The potentiometer adjusts the amount of light, from bright to dim to out, with little loss of energy since the scr controls the energy applied to the lamp. Without the scr, the energy that is diverted from the load is simply dissipated as heat across the variable resistor.

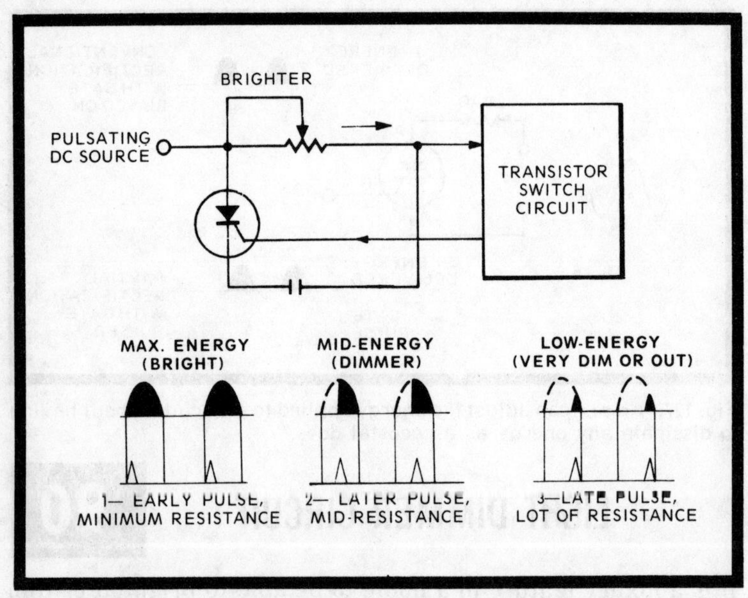

Fig. 128. An scr lamp dimmer delivers energy to a load during an interval determined by a timing pulse.

FLASHER CIRCUIT 91

A flasher circuit is used to warn motorists at night of highway construction and other such uses. The familiar yellow winking light is the output of the flasher circuit (Fig. 129). The light is turned on and off as a result of a filter capacitor's charge and discharge. The filter charges at a speed proportional to the amount of resistance in series with it. A potentiometer is one of the resistances. The setting of it determines the frequency of the light pulses.

As the switch is closed, a rectified dc is applied to the load and an scr is series connected. The RC network, across the scr, feeds into a 2-stage transistorized amplifier coupled to the gate of the scr. The filter begins charging. When the charge gets high enough, it turns on the amplifiers. The amplifiers gate the scr into conduction. The lamp lights up and stays lit until the filter discharges. When the filter discharges, the input to the amplifiers drops and they turn back off. The light does not go on again until the scr is triggered back on. This happens the next time the filter charges again.

The filter charges through the potentiometer. Therefore, that RC time constant determines the off time. Varying the pot increases or decreases the flashing rate. The filter discharges through the base input resistor. That RC time constant

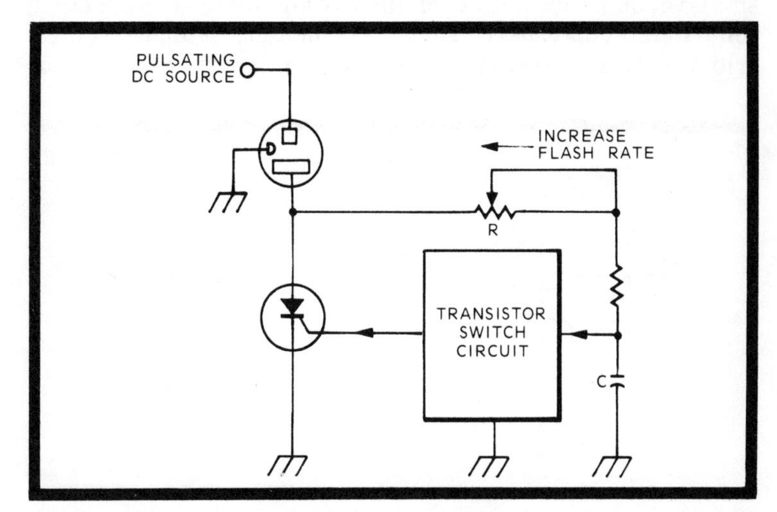

Fig. 129. An RC time constant determines the rate of an scr flasher circuit.

determines the on time. It is usually fixed at one-half second. If you want to vary it, change the value of the resistor.

92 MODEL CAR SPEED CIRCUIT

The speed of model cars, trains, etc., can be controlled with an scr circuit. Without the scr in the circuit, the model gets all the output voltage and runs at top speed. When the scr is introduced, the spped is reduced as it conducts. If it conducts all the time, the car will stop. If there is delay in triggering the scr, the amount of delay is proportional to the speed. The delay time for triggering is controlled by a potentiometer.

When the circuit is turned on, the filter charges and a voltage exists across the filter and thus the output terminals. The amount of time the scr conducts raises and lowers the dc level on the filter. The potentiometer determines the scr conduction time.

93 CRT GUN STRUCTURE

There are many types of cathode-ray tubes, but the most common are those used in TV. If you understand these, you can apply the same knowledge to others. The heart of the picture tube is the electron gun. It is like the structure in a small vacuum tube and, as of this writing, there is no practical solid-state counterpart. There is a heater, cathode, control grid, and screen grid (Fig. 130). In addition, there is a focus

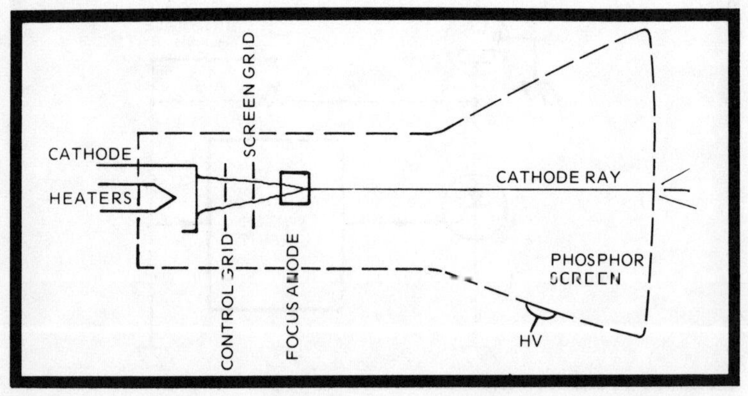

Fig. 130. A crt electron gun is like an ordinary pentode, except the plate is the phosphor screen.

grid. There is no plate as such, but there is a target—the phosphor screen.

In a black-and-white type crt there is a single electron gun. In a color crt there are, of course, three guns. The filament heats the cathode and a space cloud of electrons forms around the cathode. The control grid performs the usual valve action, depending on its electrostatic charge or bias. The screen grid accelerates the electrons toward the target. The focus grid squeezes the electrons into a tight beam. The electrons then speed toward the target, attracted by the crt anode voltage. The anode is a glass or metal envelope, which has been evacuated. As electrons collide with the target, the phosphor lights where the negative charges hit. The electrons are then collected by the anode and drained off to the power supply.

Crt Gun Failure

The main reason for crt failure is a defect in the gun, such as reduced emission, a short between elements, or an open element. When a crt gun is suspected as bad, a crt tester-rejuvenator is used to analyze the exact trouble.

A heater-to-cathode short in a picture tube can be rendered harmless by installing an isolation transformer between the heater and cathode circuits. A cathode-to-control-grid, or control-grid-to-screen-grid short can be removed most of the time by applying a shock between the elements. The shock voltage can be derived from the crt tester or the loaded capacitor trick described earlier can be used. Open elements in the gun can sometimes be welded together by vibrating the gun and shocking it at the same time. Rejuvenation is also provided by the crt tester and by adding a stepup transformer to put a higher ac voltage on the heater.

94 SHADOW MASK STRUCTURE

The conventional 3-gun color crt has a shadow mask located near the phosphor screen, between the guns and the screen. There are three sets of phosphor arranged on the screen in the form of dots, elipses, stripes, or what have you. There are red, green, and blue sets.

The dots are arranged in triads, each with one red, one green, and one blue dot. Each of the three guns is tilted slightly so that the red gun aims at the red dots, the green gun at the green dots, and the blue gun at the blue dots.

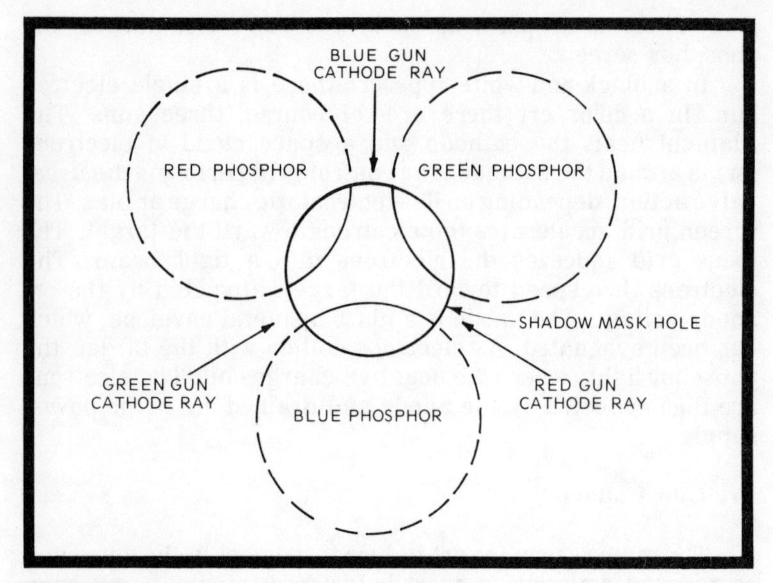

Fig. 131. The common color crt shadow mask has one hole for every three phosphor dots.

The shadow mask, placed in front of the dots, has one hole for every three dots (Fig. 131). The tilt of the electron guns makes the three beams meet, or converge, at each hole in the mask. That way all three beams pass through the same hole and each beam strikes the proper dot. About 80 percent of the electrons do not get through the mask, but bounce off and are collected by the anode voltage.

Shadow Mask Failure

The precise positioning of the mask over the phosphor is essential. The mask sometimes warps or slips. If it does, the positioning of the mask is lost and polka dots appear on the TV screen. The only repair is a new color crt.

95 DEFLECTION YOKE CIRCUIT

As the cathode ray leaves the gun, it is aimed for the center of the target. This puts a bright dot of light in the center of the screen. The cathode ray must be grasped by an electromagnetic or electrostatic field and scanned across the crt face to produce an image. In TV the cathode ray must be scanned horizontally at 15,750 Hz and vertically at 60 Hz to coincide with the transmitted TV signal to reproduce the TV picture. This is accomplished in TVs with a deflection yoke. (In oscilloscopes, plates inside the crt envelope do the job with electrostatic charges.)

The deflection yoke has a pair of horizontal coils and a pair of vertical coils. The two sets of coils are in shunt with the horizontal output and vertical output transformers. The horizontal pair is in series with each other and each has a shunt resistance across it to adjust the Q of the coil and slope the wave slightly. The vertical pair also is in series, and shunt resistances and capacitances are connected across each coil for final wave shaping. The two coils, although in the same unit, have no circuitry between them. They do independent jobs (Fig. 132).

The final waveshapes, even though they are 15,750 Hz for horizontal and 60 Hz for vertical, are both linear sawtooths. The horizontal waveshape sweeps the cathode ray from left to right slowly, then retraces it quickly back to the left side of the screen. The vertical electromagnetic energy moves the beam slowly down across the screen while it is being swept back and forth, then snaps it back up to the top. There are 525 lines during each full scan.

Yoke Failure

The yoke can be the cause of scanning troubles. Should one of the four sides of the picture pull in, giving the picture a

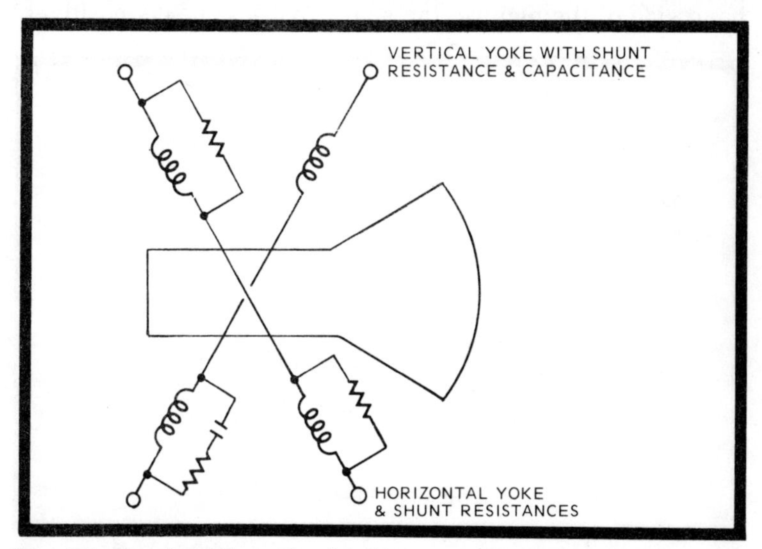

Fig. 132. The deflection yoke develops magnetic fields that deflect the cathode ray to scan the crt screen.

keystone shape, odds are that a yoke coil has shorted. Yoke failure could also cause a loss of high voltage or other types of sweep failure.

96 HORIZONTAL CONVERGENCE CIRCUIT

The guns of a color crt are tilted so that the three cathode rays meet at each hole in the shadow mask. However, as the cathode rays are swept by the deflection yoke away from the center of the screen, the three cathode rays cannot converge. The distance from each gun to a common shadow mask hole becomes different for each cathode ray. Out at the edges of the phosphor screen, the beam distances are considerable (Fig. 133). Therefore, the beams have to be spread a bit as they are swept away from center to make them converge at the shadow mask holes. To do the job there are horizontal and vertical convergence circuits.

A convergence yoke is placed around the neck of the picture tube behind the deflection yoke, between the guns and deflection yoke, and there is a horizontal convergence coil for each of the three guns. A sine wave of current is placed in each coil. As the beam is scanned from left to right, the sine wave travels 360 degrees.

When the horizontal dynamic convergence signal is in phase with the TV signal, the following occurs: At the left-hand side of the picture, the sine wave just peaks positively

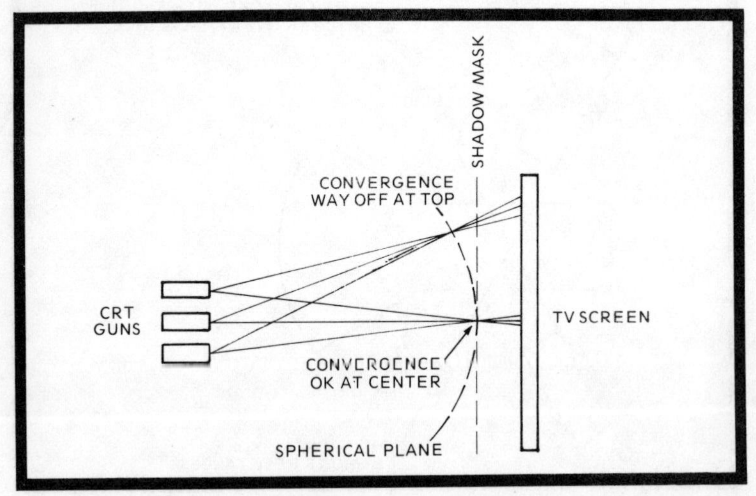

Fig. 133. The convergence point of the three beams in a color crt varies because the faceplate is not spherical.

and is heading downwards. This puts a maximum convergence voltage on the coils mounted over the guns. This voltage presses the cathode rays. Since there are adjustable inductance and resistance controls on each coil, the beams can be adjusted to converge at the shadow mask.

As the scan approaches screen center, the sine wave approaches the most negative point of its curve. This puts a minimum convergence voltage on the coils; this also is adjustable. Thus, the cathode rays can be made to converge at the center of the shadow mask. As the scan approaches the right side of the screen, the sine wave approaches the next positive peak. This puts a strong voltage on the coils again, permitting right-side convergence. As the sync pedestal reaches the gun, the screen is blanked out for the retrace. The sine waves of current are taken from the horizontal output transformer and modified with RC and LC networks for convergence coil application.

There are six controls on the horizontal convergence coils—three for adjusting the sine-wave amplitude and three for adjustment of the sine-wave phase. This changes the waveforms enough to adjust the convergence. The horizontal convergence procedure for each receiver is listed in the manufacturer's service manual. For horizontal convergence troubles, localization is usually easy. Simply try adjusting the different coils. The control circuit that doesn't work is defective.

VERTICAL CONVERGENCE CIRCUIT 97

While the horizontal convergence coils are fed a sine wave, the vertical coils need a parabolic wave (Fig. 134). Actually, though, if you examine the part of the sine wave that actually affects the cathode rays during horizontal convergence, it is a parabolic wave (Fig. 135). The rest of the sine wave occurs during the blanked-out retrace time.

The vertical driving pulse is taken from the vertical output transformer and is a 60 Hz sawtooth. It is passed into an integration circuit that changes the sawtooth to a parabola. The parabolic waveforms are injected into vertical convergence coils. This waveform also presses against the cathode ray, but at a 60 Hz rate to affect the vertical sweep. The parabola is at maximum voltage at the top of the picture, minimum voltage at the center of the picture and then rides back to maximum voltage at the bottom of the picture.

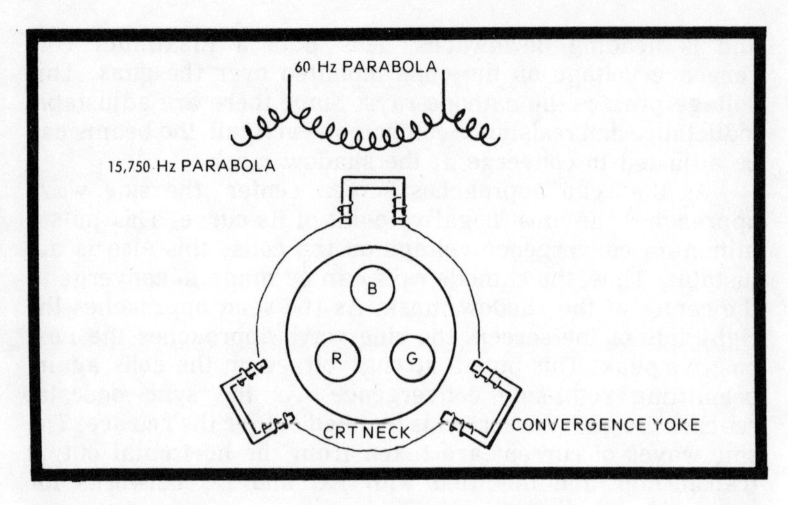

Fig. 134. The convergence waveform consists of two parabolas merged together.

The adjustments enable the servicer to adjust the amplitude and phase of the voltage to maintain convergence of the cathode rays during the entire scan. Both the horizontal and vertical convergence circuits are critical and when replacing components actual factory-recommended parts should be used. Using another manufacturer's diode can cause incorrect amplitudes or phase and make a repair job difficult.

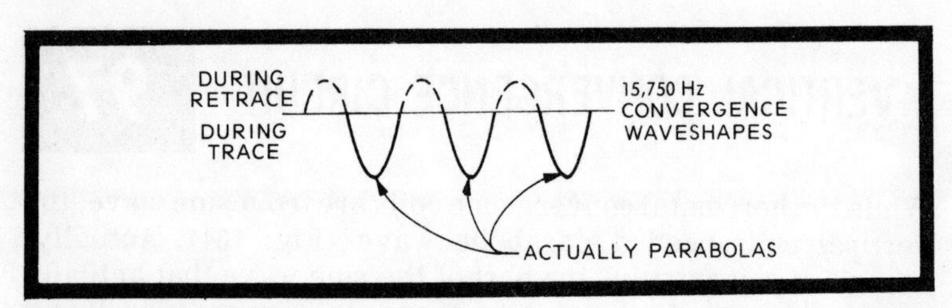

Fig. 135. A sine wave can be used as a parabola waveform if applied as shown.

98 and 99 VERTICAL SWEEP

The vertical sweep section of a TV produces a sawtooth waveform which is applied to the vertical deflection coils. In color TVs it also sends a signal to the convergence coil circuits. In a tube TV, the sweep circuit consists of an oscillator

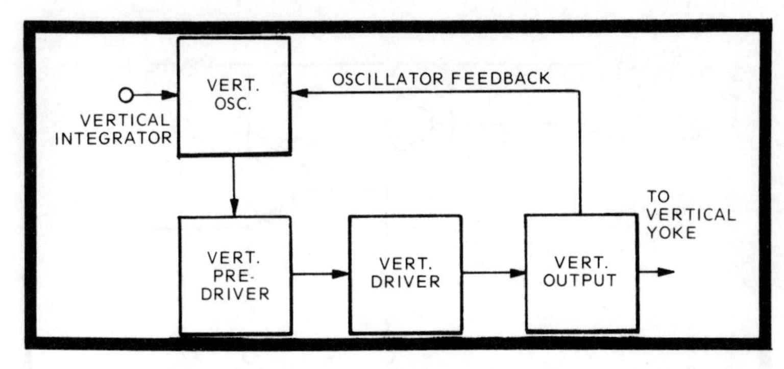

Fig. 136. A solid-state vertical sweep circuit is similar to the tube counterpart, except for the two driver stages.

and output. In solid-state, the same two circuits are there, but in between is a predriver and a driver (Fig. 136).

A typical solid-state configuration shows four pnp transistors in parallel. The first one is an oscillator (Fig. 137). It runs at a 60 Hz rate and is synchronized to the transmitter by a pulse from the vertical integrator. The predriver and driver are tow pnp types in parallel with the collectors tied together (Fig. 138). The predriver acts as an emitter follower from the driver, and the driver acts as an emitter follower for the output. That way they prevent interaction between the oscillator and output and match the 60 Hz wave into the low impedance of the output.

There is a feedback network from the output to the oscillator. This makes the output circuit part of the oscillator. The feedback waveform is a modified sawtooth and is taken just as the yoke receives the driving signal.

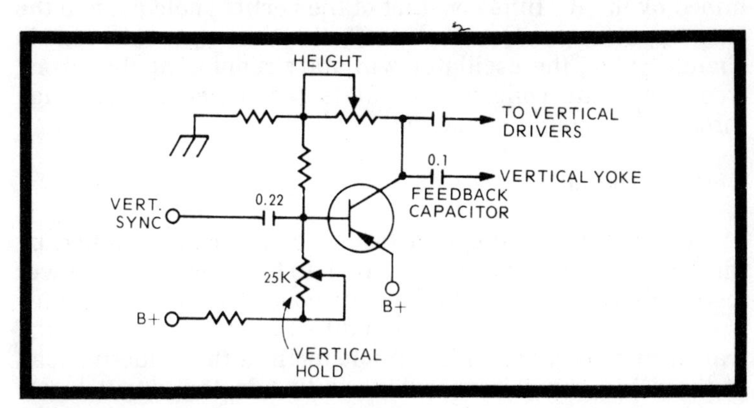

Fig. 137. The typical vertical oscillator circuit gets its feedback from the vertical yoke coils.

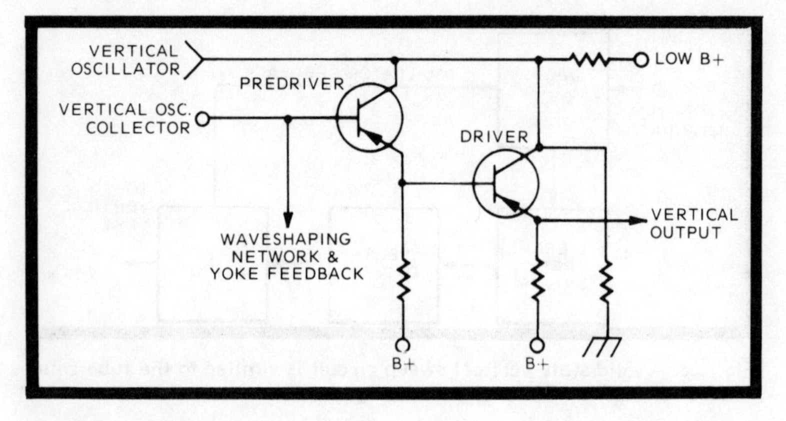

Fig. 138. The vertical driver stages feed the vertical output with an emitter follower.

Vertical Oscillator

The oscillator is an electronic switch. It turns on and off at the 60 Hz rate and oscillation is sustained by the output feedback (Fig. 137). The input sync pulse synchronizes the oscillator so it is on or closed during the vertical retrace time as the cathode ray is brought back up to the top of the scan. The switch is off or open during the vertical sweep time, as the cathode ray moves from top to bottom.

The switch is on a much shorter time than it is off. This action is controlled by the charge and discharge rate of a carefully calculated 0.1 uF feedback capacitor from the yoke to the oscillator collector. The oscillator frequency is determined by the RC time constant of the vertical hold pot and the base input capacitor. If one of these components should change value, the oscillator will start running at the wrong frequency. The collector output is fed to the driver stage through a coupling capacitor.

Vertical Output

The sweep signal output from the driver is amplified by the vertical output transistor (Fig. 139) to a high enough power level to drive the vertical deflection coils. An autotransformer is in the collector circuit. The modified sawtooth output of the transformer produces a linear sawtooth in the inductive load of the yoke. There is an extra winding in the output transformer for feedback to the oscillator base to produce the oscillation.

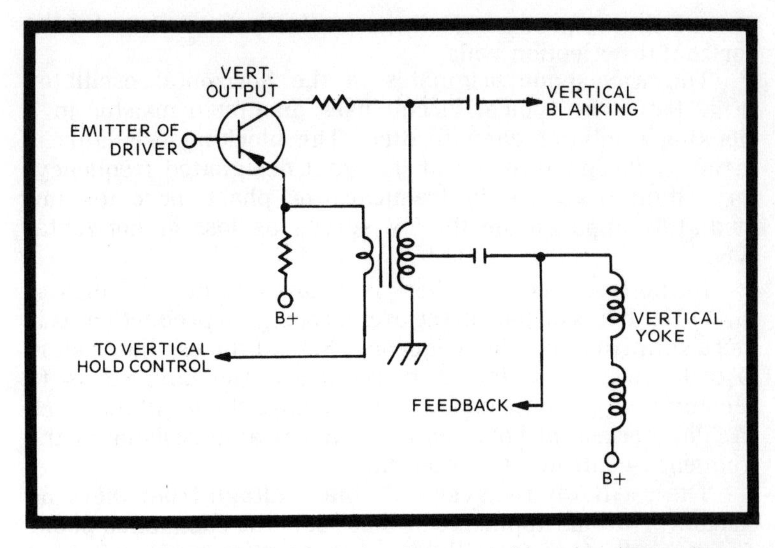

Fig. 139. The vertical output circuit also provides a blanking pulse.

HORIZONTAL OSCILLATOR 100

In addition to scanning chores, the horizontal sweep system has the job of producing the crt anode accelerating voltage.It

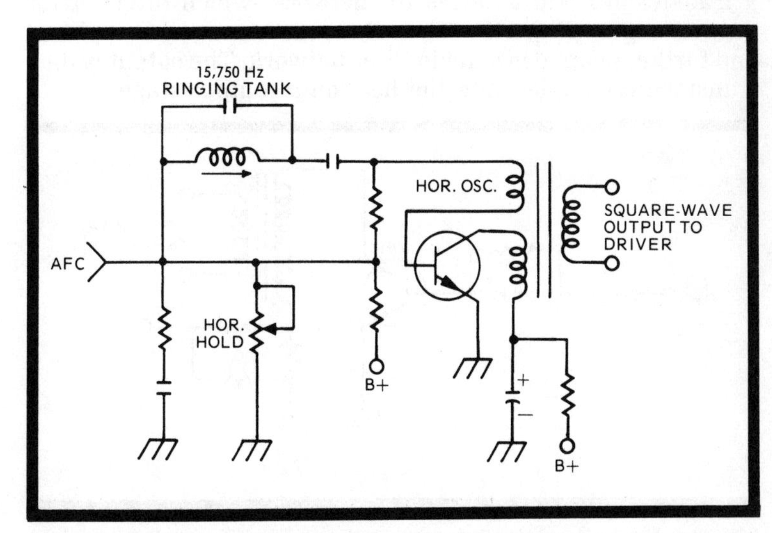

Fig. 140. The blocking-type horizontal oscillator produces a square-wave output.

does all this by producing a large sawtooth waveshape for the horizontal deflection coils.

The waveshpae originates in the horizontal oscillator (Fig. 140). The typical circuit uses an npn transistor in a blocking oscillator configuration. The blocking oscillator is carefully designed to run at its exact designated frequency. Any slight deviation in frequency or phase becomes immediately apparent on the TV screen as loss of horizontal sync.

In the feedback line, there is a tank circuit that rings at 15,750 Hz. It has no part in the oscillator signal production. It is there simply to hold the frequency. Should the oscillator start to drift away from the tank frequency, the tank starts to become a larger impedance, which brings the oscillator back on. The horizontal hold control is a variable resistor in the frequency-controlling components.

The oscillator receives a dc bias voltage from the sync circuits. The bias applied to the base sets the conduction point. The output of the oscillator, due to the good switching characteristics, is a square wave.

101 HORIZONTAL DRIVER

The horizontal driver is not found in tube receivers, only solid-state types. The driver takes the square-wave output through a transformer and a series RC network, which filters it for ragged edges (Fig. 141). Then the square wave is amplified and further shaped in a diode-filter network. The output is then transformer-coupled into the horizontal output stage.

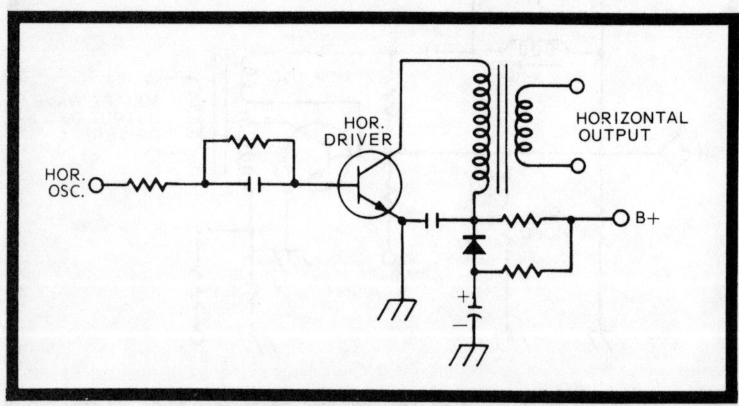

Fig. 141. The horizontal driver passes the square wave into the output through a matching transformer.

HORIZONTAL OUTPUT

The horizontal output transistor receives the square wave across the base-emitter junction. The collector is grounded and the load is in the emitter circuit (Fig. 142). The desired output is power, and in an emitter follower, a large current gain is obtained. Connected to the emitter are the flyback transformer, horizontal deflection coils, and the damper diode. The damper cathode is attached to the emitter and its anode is grounded. Any other coils or components that might be found in the circuit are there to help the damper by dampening any parasitic oscillations that might get started due to the high frequency and large power requirements.

The modified square wave turns on the output transistor in the center of its linear rise while the electron beam is in the center of the TV screen (Fig. 143). The emitter current rises linearly for the rest of the sweep until the scan reaches the right side of the picture. At the right side of the picture, the driving waveform rises suddenly. This turns the pnp off. This causes the yoke current to disappear. At this instant the magnetic field suddenly collapses. The speed of the collapsing field produces a large voltage in the yoke that is coupled into the flyback transformer. This large voltage is rectified and filtered by the high-voltage rectifier and fed to the crt anode.

When the magnetic field collapses the cathode ray quickly retraces to the left side of the picture. The light is blanked out by a pulse added to the electron gun during retrace. (This is discussed in the blanking section.)

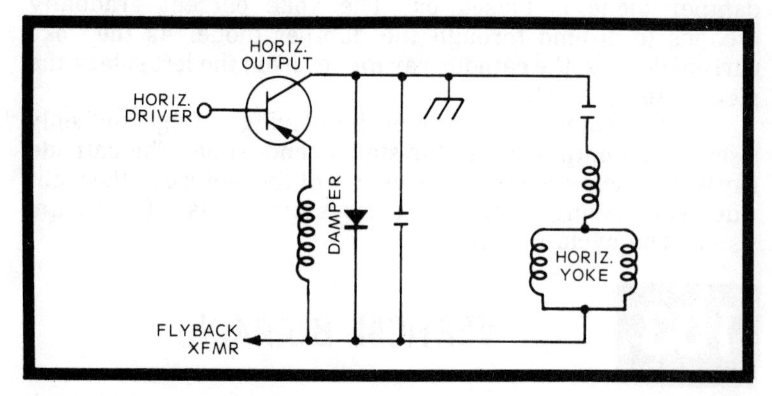

Fig. 142. The horizontal output sends a pulse to the flyback and drives the horizontal yoke.

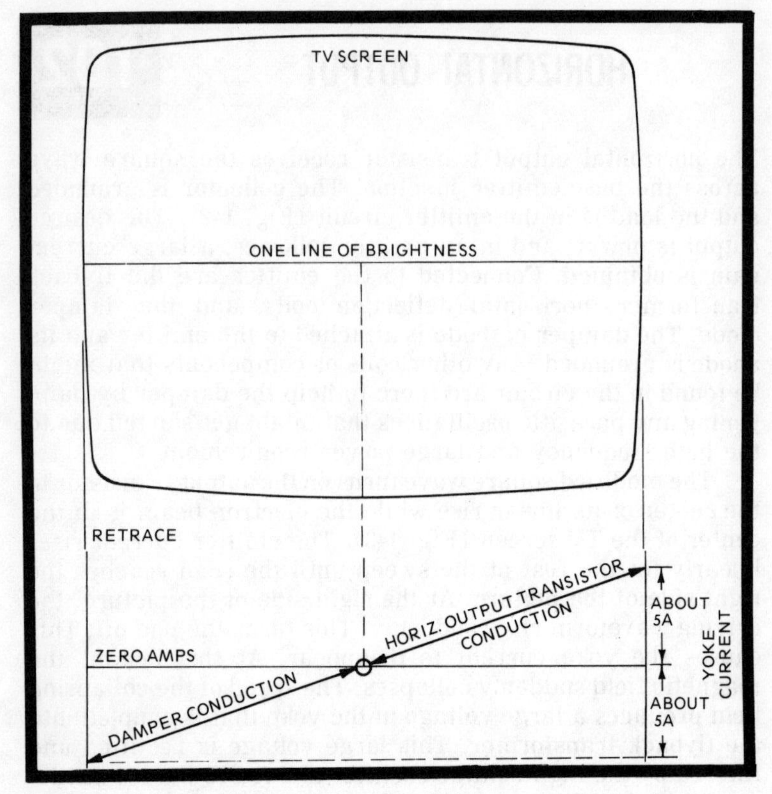

Fig. 143. Various current waveforms existing in the horizontal output during one line of horizontal sweep.

At the end of the retrace, the waveshape drops low and the damper diode is biased on. The yoke current gradually escapes to ground through the damper diode. As the yoke current decays, the cathode ray moves from the left side of the picture to the center.

At the center of the screen the driving voltage suddenly rises as the output transistor starts conducting. The cathode ray is then deflected from the center of the picture to the right side. The driving voltage then rises and turns off the transistor. The cycle repeats.

103 VERTICAL BLANKING

As the TV picture is being scanned vertically, the beam has to be on during the scan and extinguished during the retrace

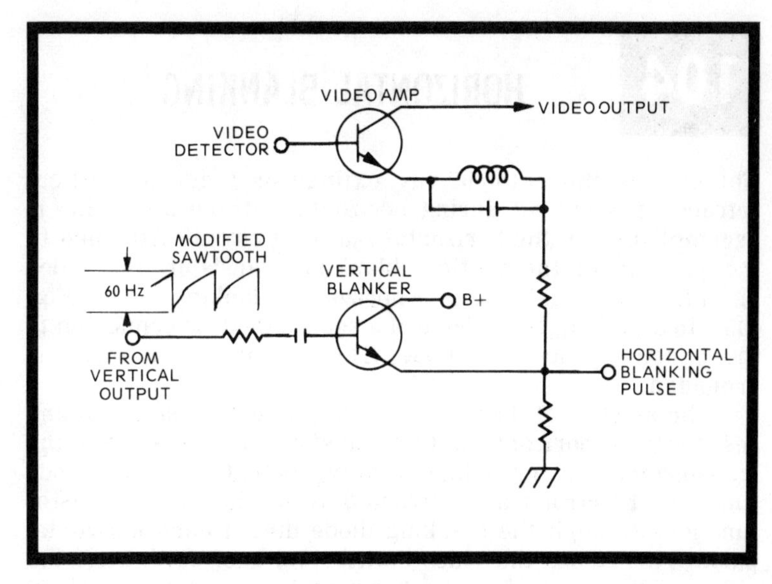

Fig. 144. The vertical blanking pulse causes the video amplifier to cut off during the vertical retrace time.

from the bottom of the picture back up to the top. There are 525 lines in each complete picture frame and 30 frames are televised per second. There are two fields for each frame—the odd field and the even field. First one is scanned and then the other. That makes two retrace intervals for each frame. Therefore, a vertical blanking pulse is transmitted 60 times every second. As the vertical blanking pulse arrives at the crt cathode, it biases the crt beyond cutoff.

However, TV manufacturers are not completely satisfied with this blanking procedure, so they add vertical blanking circuits to strengthen the effect of the transmitted signal. The vertical blanking circuit consists of a vertical blanking transistor and an input from the vertical output circuit. A resistor and capacitor in series is attached to the top of the vertical output transformer. It picks up the modified sawtooth and blocks the dc.

The pulse is fed to the base of an npn transistor (Fig. 144). The emitter of the npn is connected to the emitter of the video amplifier. As the vertical pulse goes positive, the transistor conducts. This raises the voltage on the emitter of the video amplifier, also an npn. The increased potential cuts off the video amplifier. The vertical pulse, therefore, cuts off the video and blanks out the crt. Since the vertical pulse occurs during retrace, blanking occurs at the correct time.

104 HORIZONTAL BLANKING

Not only is the cathode ray extinguished during vertical retrace, it is cut off during horizontal retrace also. This is accomplished by the horizontal blanking circuit. Attached to the junction of the vertical blanking transistor and video amplifier is the cathode of a horizontal blanking diode (Fig. 145). In addition, the cathode of a second diode is connected to the anode of the first diode. The second diode anode is grounded.

The junction of the two diodes is connected via a coupling resistor to the horizontal output transformer. A pulse from the transformer creates a high positive potential at the diode junction. Electrons are attracted from the two transistor junctions through the blanking diode during each horizontal pulse. This raises the video emitter voltage, cutting off the video amplifier during the horizontal retrace time. Thus, horizontal blanking is accomplished.

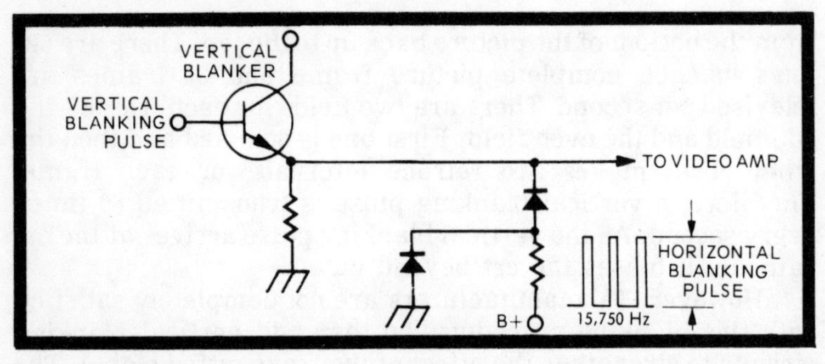

Fig. 145. The horizontal blanking pulse causes the video amplifier to cut off during the horizontal retrace time.

105 DC RESTORER

Dc restoration is a common technique, yet is a largely ignored circuit by servicers. Actually, it is a simple circuit. It is needed when the dc level of a circuit is lost as the signal passes through blocking capacitors. The ac goes through fine but the dc level remains behind.

A diode with a resistor across it is used to restore the dc level. A typical use of a dc restorer is in the crt cathode input

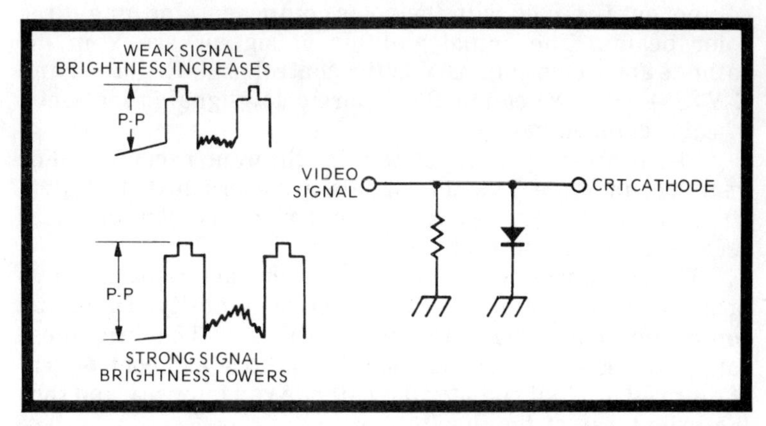

Fig. 146. A dc restorer keeps the dc bias voltage on the crt cathode at the same level, no matter the signal strength.

circuit (Fig. 146). This is because with an ordinary grid-leak resistor setup, different levels of dc bias are developed as incoming signals vary in strength.

As the signal passes over the anode of the diode, it causes electrons to come from ground and then go back to ground through the parallel resistor. If the signal is weak, little current flows. If the signal is strong, a lot of current flows. The amount of bias voltage on the anode varies according to the amount of signal.

The dc level on the crt cathode determines the brightness level. Therefore, when the signal is strong, the dc level rises positively, reducing the brightness level. As the signal becomes weak, the dc level drops, thus increasing the brightness level. This is what is needed since the grid capacitor develops a charge in accordance to the strength of the signal. The diode keeps the charge at the same dc level.

COLOR TV RECEPTION

A reproduced color TV picture is actually four separate pictures, one on top of each other. There is the black-and-white part, then there are red, green, and blue components. The black-and-white picture is the same as the one in a monochrome TV. It is processed by the video amplifiers (called the Y channel in a color receiver) and is applied to all three cathodes of the three guns in the color crt.

Color signals are applied to the control grids of the guns, the red signal to the red gun and so forth. The black-and-white component is mixed with the color component, forming three color beams. The actual addition of signals are Y in the cathode and a color minus Y in the control grid, R-Y, B-Y, and G-Y(Y + (R — Y) equals R). A pure color signal is contained in each cathode ray.

There are variations of this in different receivers. For instance, in some TVs the signals are combined in matrix circuits and simply fed to the control grids. The cathodes receive no signal in that type of circuit.

The displayed colors have three characteristics. One is brightness, two is tint and three is color intensity. There is a control for each characteristic in the TV. The brightness control adjusts the bias between the cathode and control grid of the crt gun. That changes the cathode ray intensity and thus the brightness of the display.

The tint control changes the phase of the color oscillator. This produces different colors. The flesh tones range typically from pink through normal to green.

The color intensity control changes the amount of color amplification in the color i-f stages. This produces more or less vividness in the colors on display. It seems to change the color, but it's only the color intensity. It ranges from no color, to normal flesh tones, to vivid orange flesh tones.

The color information is contained in the composite color TV signal, which is made up of two different types of signal. One is the color sideband energy that contains the color information. The sidebands occupy spectrum space on either side of the 3.58 MHz color subcarrier. Two is the color sync signal called the burst. It is located on the back porch of the horizontal sync pedestal. It consists of eight or nine sine wave rings. The ringing is designed to have the precise frequency and phase needed to lock the receiver's 3.58 MHz oscillator in step with the incoming signal. The output of the video detector contains the sidebands and burst. Signal is fed from the video detector to both the color i-f stages and the burst amplifier.

106 COLOR I-F CIRCUIT

A sample of the detector output arrives at the color i-f input. In the video sample are the color sidebands on either side of 3.58 MHz. The color i-f can be a single- or a double-stage circuit. The input is capacitively coupled and the complete signal enters the base or control grid (Fig. 147). The signal output is developed across a burst transformer. The burst transformer is tuned around the 3.58 MHz range. It develops the strongest

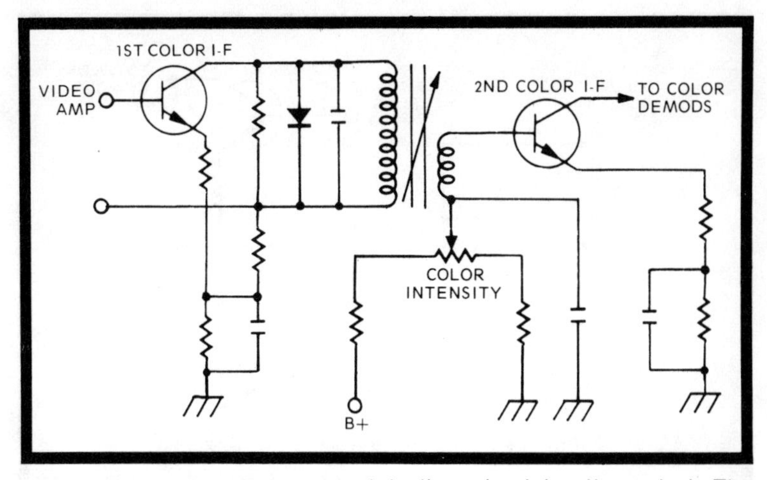

Fig. 147. The color i-f stages contain the color intensity control. The transformer is tuned to 3.58 MHz.

output there. As a result, the rest of the signal containing video and audio is rejected.

It is also desirable to get rid of the horizontal sync pulse containing the burst information, since it is ringing at a 3.58 MHz rate and would interfere with the sideband amplification. A lead from the flyback transformer is attached to the emitter or some convenient point. During the flyback retrace, the color i-f amplfier is cut off. That way the sidebands are amplified intact and appear across the load.

COLOR KILLER CIRCUIT 107

The color killer circuit is needed when no color is being received. If the color i-f stages were amplifying during a monochrome transmission, noise would be amplified and cause color streaking or snow in the picture. The color killer biases the color i-f stages off during a monochrome transmission and turns them on during a color program.

The color killer, typically a transistor or tube, is operated by the color phase detector (Fig. 148). When the detector is processing the burst, it provides a negative voltage to the base or control grid and biases the color killer off. With the color killer not conducting, no voltage is developed across its load resistor; the load resistor is designed to be a part of the input bias circuit of the color i-f. When there is no voltage across the resistor, the color i-f amplifies normally (Fig. 148).

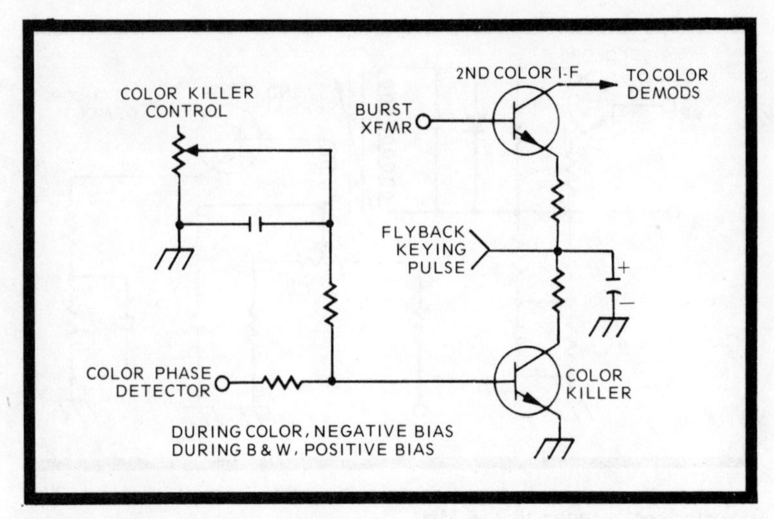

Fig. 148. The color killer does not affect the color i-f stages during color programing but shuts down during black-and-white reception.

During a black-and-white program there is no color burst. As a result the color phase detector does not conduct and provide the killer with its negative voltage. The color killer, without its bias, turns on. A voltage appears across its load, which is also the color i-f bias, and the color i-f stage turns off. This shuts down the color circuits and there is no chance of them interfering with the monochrome transmission.

The color killer is an amplifier. As such it can and does pick up noise and amplify it. This noise could be coupled into the color i-f, due to the mutual resistor. In order to avert such a happening, the color killer is allowed to conduct only during the horizontal sync pulse interval. This is accomplished by keying the killer with a pulse from the flyback. The pulse is attached to the collector or plate of the killer. Normally, there is no B-plus on the plate. Therefore, the only time the killer can conduct is when the plate receives the high positive pulse. As a result of these limited conduction intervals, the killer remains noise free.

There is a killer control on every color TV, and the adjustment is critical. The control should be set on an off channel. Turn it until you get a lot of colored snow. Then reverse the control until you just get rid of the color and have black-and-white snow. This is the best bias spot. With abnormal signals, such as too strong or too weak, this setting might have to be slightly changed to compensate for the abnormality.

BURST AMPLIFIER CIRCUIT 108

The burst amplifier has the job of extracting the burst rings intact from the horizontal sync pedestal. These rings contain the frequency and phase information the color oscillator must have in order to faithfully reproduce the color picture the transmitter is sending out over the air.

The burst amplifier is reverse biased or cut off if it is a tube. Attached to the emitter or cathode is another line from the versatile flyback transformer (Fig. 149). The gating pulse turns on the burst amplifier during the flyback retrace. The burst amplifier conducts as the sync pedestal enters its input. It is turned off during the video interval. Only the burst rings appear in the output. The burst is developed across the burst output transformer. The transformer is a specially wound and tuned, tightly coupled type.

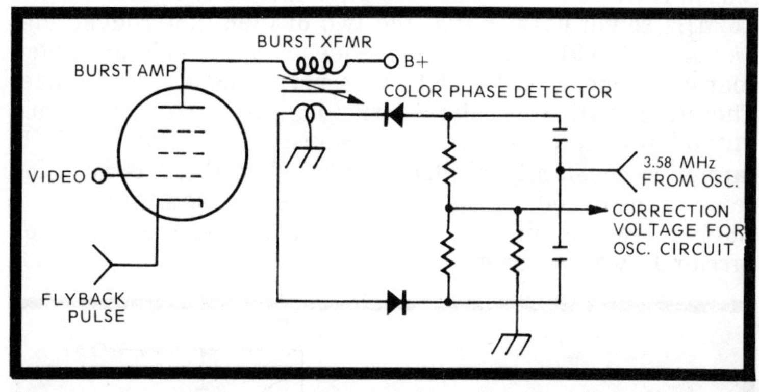

Fig. 149. The burst amplifier feeds the color phase dectector where burst is compared with the 3.58 MHz signal from the local oscillator.

COLOR PHASE DETECTOR CIRCUIT 109

The phase detector is a dual diode circuit (Fig. 149). The secondary of the burst transformer is attached directly to the cathode of one diode and the anode of the other. The center tap of the transformer is grounded. The burst output passes from one diode to the other through a pair of load resistors and back to the transformer. Equal and opposite voltages are developed at the top and bottom of the two resistors.

In parallel with the load resistors are two capacitors in series. At their center a second signal is applied. This is the free-running CW from the 3.58 MHz color oscillator. Since the burst is also 3.58 MHz, the two signals are compared in the detector. If they are both identical they simply add and equal and opposite voltages still exist at the top and bottom of the load resistors. Should the frequency and phase not be identical, the plus and minus voltages at the end of the load will shift. Then, the center tap at the load, instead of being 0V, will develop a correction voltage. This voltage is fed back to the color oscillator to compensate for the deviation and lock it in step with the burst.

110 COLOR OSCILLATOR CIRCUIT

Before the color signals are broadcast, a 3.58 MHz filter removes the color carrier, leaving only the sidebands. The reason is twofold. One, the 3.58 MHz carrier is so close to the 4.5 MHz sound carrier that the two of them heterodyne and produce a 920 kHz difference frequency beat. This unwanted beat can appear in the picture as herringbone interference. Therefore, with the color carrier suppressed at the transmitter, this interference is lessened. Two, the 3.58 MHz carrier is constantly produced even when there is no color being transmitted. If the carrier is suppressed, and only a black-and-white picture is transmitted, there is no color carrier to worry about.

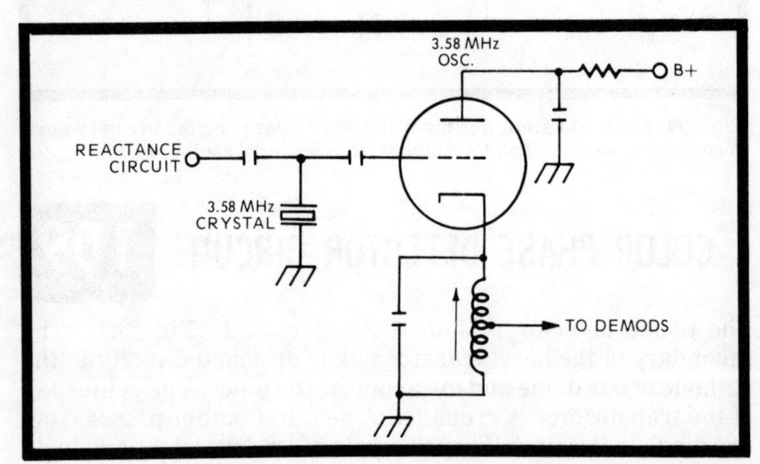

Fig. 150. The color oscillator is a typical crystal-controlled circuit operating at 3.58 MHz.

However, the color carrier suppression at the transmitter requires an additional function in a color receiver. Before the color sidebands can be processed, the color carrier must be regenerated and mixed with the sidebands. This is accomplished by having a 3.58 MHz oscillator in the color circuitry, which is kept in step with the color subcarrier at the TV station by the burst.

In many color TVs there is a reactance circuit across the oscillator. The reactance acts as a variable capacitance or inductance according to the dc correction voltage applied from the phase detector. The typical color oscillator (Fig. 150) is a crystal-controlled type with a variable tank in the emitter or cathode.

COLOR DEMODULATOR CIRCUITS

The color demodulator has a double job. First it must mix the 3.58 MHz CW signal coming from the oscillator with the amplified sideband coming from the color i-f (Fig. 151). Once they are mixed properly the original color carrier modulated with the red, green, and blue signal components is restored.

The mixing takes place in the transistor or tube. There are two or three demodulators, according to the method selected by the manufacturer, but they do a similar job. When there are two, one usually processes red and the other blue. The red demodulator gets sidebands directly from the color i-f and a 3.58 MHz signal directly from the oscillator. One of the signals enters through the base or control grid, while the second enters through the emitter or cathode. They mix in the electron

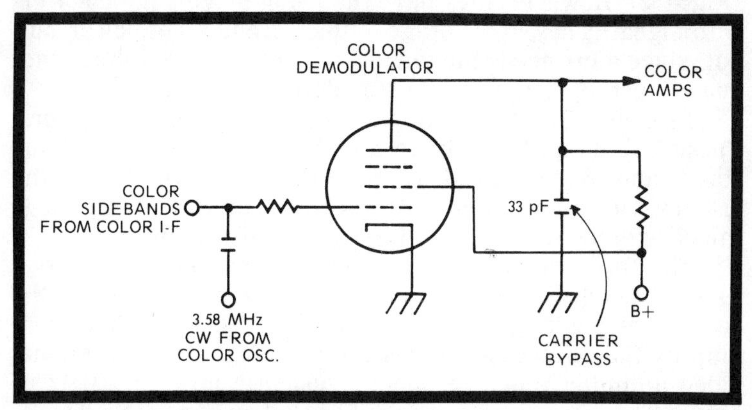

Fig. 151. The color demodulator mixes the sidebands and CW, then detects the color signal.

flow and a load resistor or coil develops the output, which is the R-Y signal modulating the subcarrier.

The blue demodulator gets sidebands directly from the color i-f but receives CW after it has been passed through a phase-shift network. The network is a simple RCL setup that changes the phasing according to manufacturer specifications. It could be 57.5 degrees or 90 degrees or what have you. The phase change causes the amplifier output to have B-Y modulating the 3.58 MHz subcarrier. The G-Y input is developed by adding R-Y and B-Y together across a resistor in the G-Y amplifier input.

The second job the demodulator circuit does is get rid of the 3.58 MHz carrier. This is accomplished easily with a small capacitor to ground in the collector or plate circuit. The 3.58 MHz signal is bypassed while the lower frequency R-Y and B-Y signals pass right over the bypass. An rf trap tuned to 3.58 MHz is sometimes used instead of the capacitor. Either way, the 3.58 MHz finds a low-impedance path to ground while the color-difference signals, with a frequency near 2 MHz, see a high-impedance path.

Some TVs have three demodulators and thus need two phase-changing networks to feed two of them with a subcarrier signal of the correct phase.

112 COLOR-DIFFERENCE CIRCUITS

Color-difference amplifiers are actually the color output stages which amplify R-Y, B-Y, and G-Y difference signals (Fig. 152). However, they have the job of driving the color crt with signal. They are voltage output circuits, not power output, since a crt needs high levels of peak-to-peak voltage, not heavy current like a speaker or deflection coil.

In some TVs there are no color-difference amplifiers. These TVs do all the voltage amplifying in the demodulator stages and send pure red, green, and blue information to the crt, not difference signals. In these circuits the Y signal is mixed with the R-Y, B-Y, and G-Y in the color amplifier.

The three amplifiers are typically transistors or triode tubes. They are essentially video amplifiers and have all the respective peaking coils. The only difference is that they amplify the difference signals up to 1.5 MHz. A conventional video amplifier has a frequency response up to 3.5 MHz or better. The color signals need not be that rich in detail. The R-Y, B-Y, and G-Y amplifiers feed respective grids in the crt.

The Y is applied to the cathodes. Thus, matrixing (mixing) takes place in the crt.

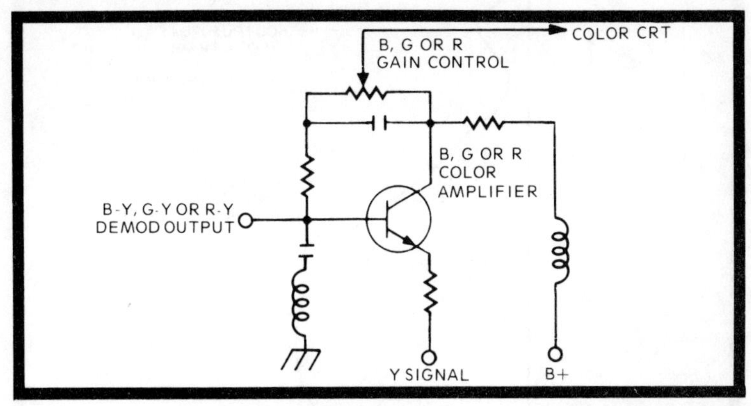

Fig. 152. A transistorized color amplifier with the Y applied to the emitter and a color —Y to the base will produce pure color in its output.

TV FIRST I-F CIRCUIT 113

The typical three i-f stages in a TV are usually considered in a group as the i-f strip. Each i-f has its own identity and must not be confused with the second or third even though between them they do a job called "stagger tuning." All that means is that the TV signal is so broad, no one i-f can amplify it alone. Each one does its own job and amplifies a portion of it.

The i-f amplifier is a typical transistor or tube. The first i-f receives an agc dc correction voltage that determines the degree of amplification. On strong signals the dc correction reduces conduction so the signal will not overload the amplifiers. On weak signals the dc increases conduction so the signal will be amplified as much as possible.

The first i-f has two traps. One trap is the adjacent-channel video. On 44 MHz i-f strips, this trap is set at 39.75 MHz. It stops any video from the next channel up from getting through and appearing in the TV screen. The second trap is adjacent-channel sound. This trap is tuned to 47.25 MHz and halts any audio from the next channel down from getting through (Fig. 153).

The other tuning section is the coupling transformer from the first i-f to the second i-f. It is tuned to the picture carrier at 45.75 MHz. It couples maximum picture carrier from the first to the second i-f. Otherwise, the first i-f is an ordinary rf-type amplifier.

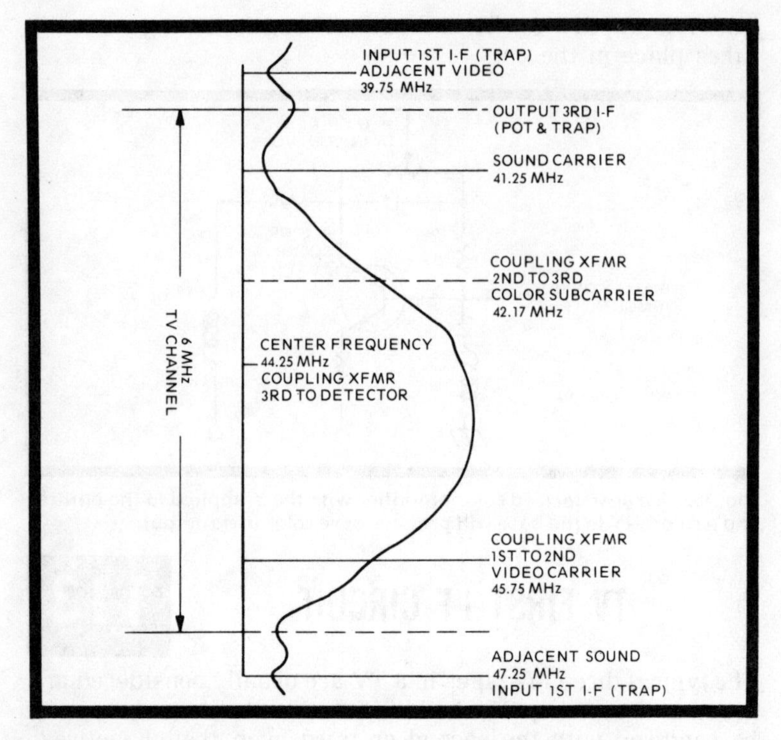

Fig. 153. Bandpass curve of an ideal TV i-f system.

114 SECOND I-F CIRCUIT

The second i-f in the strip is like the first, and quite often also receives an agc dc correction voltage. With tubes the agc is always a negative bias when applied to the control grid and positive when applied to the cathode. With transistors the agc is negative when applied to the base of an npn and positive when applied to the base of a pnp in conventional agc systems.

Typically, there is no trapping in the second i-f. The only tuned circuit is the coupling transformer from the second i-f to the third i-f. It is tuned to the color subcarrier at 42.17 MHz.

115 THIRD I-F CIRCUIT

The third i-f does not have any agc applied to it. Whatever gets through the first two i-f stages is amplified without restraint in the third i-f. In a color TV, a detector diode is connected to the output of the third i-f. It takes off a sampling of the signal for the audio detector circuit. The audio carrier at 41.25 MHz and the picture carrier at 45.75 MHz heterodyne and produce a difference carrier of 4.5 MHz. This is what is needed by the audio detector.

In the output of the third i-f, after the 4.5 MHz sound is taken off, a sound i-f carrier trap is installed. It is tuned to 41.25 MHz. The trap consists of a coil tuned to that frequency and a series potentiometer called the sound reject. Between the two they suppress the 41.25 MHz carrier. The sound carrier is 920 kHz away from the color subcarrier at 42.17 MHz. Suppressing the 41.25 MHz carrier eliminates the possibility of a 920 kHz beat. The other tuned part is the output coupling transformer. It is tuned to the middle of the bandpass at 44.25 MHz.

The three i-f transformers, being stagger tuned, reproduce the entire TV signal (Fig. 153) with the color subcarrier on one side at 42.17 MHz, the picture carrier at the other side at 45.75 MHz, and the center of the band at 44.25 MHz.

AGC-SYNC CONFIGURATION

The agc and sync circuits all interact. When servicing the general circuit area they should be considered in the same

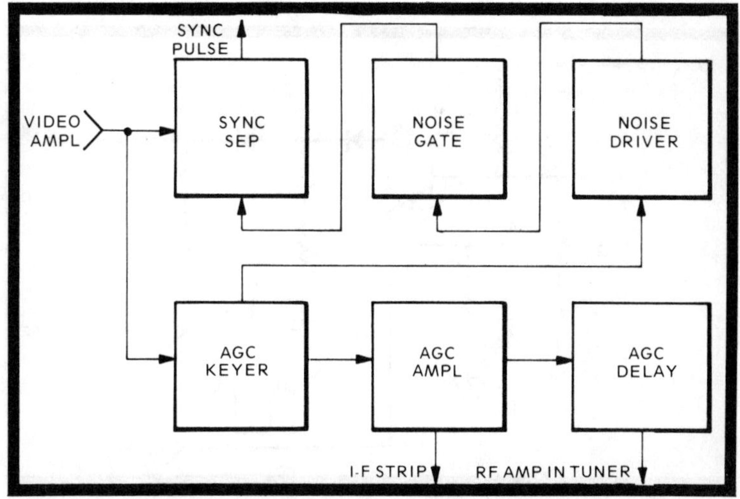

Fig. 154. The agc and sync circuits both receive similar inputs and work with the noise gate circuit.

block diagram approach. A typical setup uses six separate circuits designed around six transistors (Fig. 154). They perform three functions. One is agc, two noise immunity, and third is extraction of sync pulses for locking the vertical and horizontal oscillators in step with the transmitter.

116 AGC KEYER CIRCUIT

The agc keyer in a transistor circuit receives two inputs. One is a pulse from the flyback transformer during retrace, which is the horizontal sync pulse timing. In the circuit in Fig. 155, the pulse passes through the series resistor, capacitor, and diode. The diode rectifies the pulse and electrons pile up on the keyer collector, forming a —20V potential.

Coming in to the base is a sample of the detected video. For this pnp transistor the video is negative-going. Until the video appears, the transistor is cut off by a slight reverse bias on the EB junction. As soon as the high negative voltage of the horizontal sync pulse arrives, the base goes negative, which turns on the transistor for the short duration of the pulse. The base diode prevents any of the pulse from going into the other circuits since the diode, with its anode toward the base, becomes a very high resistance.

When the transistor turns on, electrons flow heavily from collector to base and collector to emitter. The —20V on the

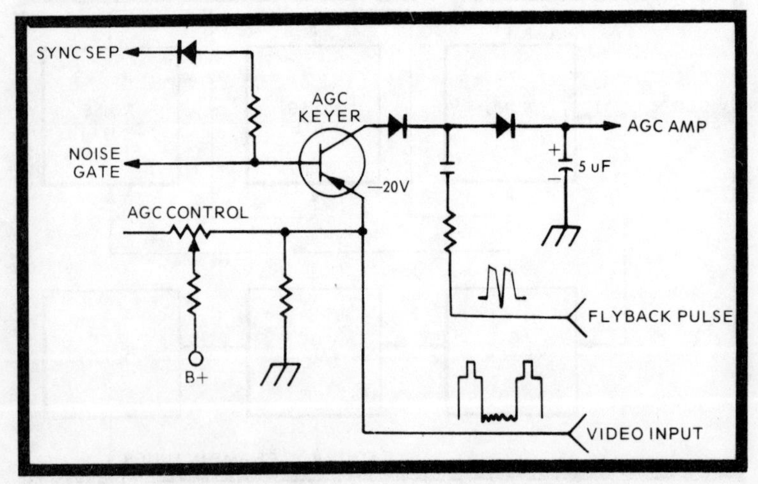

Fig. 155. An agc keyer conducts only when the horizontal sync pulse and flyback pulse arrive together.

collector drops to near zero. Electrons from the base circuit of the next stage, the agc amplifier, flow through the two diodes to the keyer collector. The agc amplifier develops about 6V positive as its electrons leave and pass to the gate. A 5 uF filter to ground charges and maintains the 6V base agc amplifier voltage between pulses.

To sum up, the horizontal sync pulse and the flyback pulse, when both are applied to the transistor, cause gate conduction. The gate is off at all other times. That way, noise or interference cannot cause the agc circuit to turn on and create noise problems.

AGC AMPLIFIER CIRCUIT 117

While there is no signal in the TV, the gate collector potential is 20V and the amplifier base is about 2V. As a signal comes in, the sync pulse drops the gate collector down near zero, pulling electrons from the amplifier base, increasing its voltage to about +6V. The 5 uF filter maintains the charge between sync pulses.

With the signal applied, the amplifier base (Fig. 156) is thus forward biased and the transistor conducts during the signal processing interval. If the signal is strong, the gate conducts heavily, making the amplifier base even more positive. Should the signal be weaker, the horizontal sync pulse is not so high, the gate doesn't conduct as much, the collector drop is lower, not as many electrons are drawn out of the base circuit, and the agc amplifier conducts less. The 5 uF filter maintains the charge caused by the sync pulse level. The charge varies plus or minus a volt or two. The output of the agc

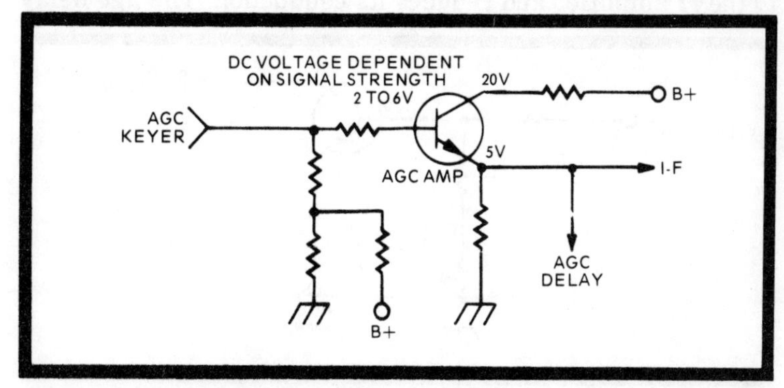

Fig. 156. The agc amplifier feeds the i-f stages directly and also sends some signal to the agc delay.

amplifier, in turn, varies according to the peak-to-peak amplitude of the sync pulse.

The agc output emitter voltage is about 5V. The collector is set near +20V. The npn conducts when the base exceeds the emitter 5V. The conduction is dependent on how positive the base goes.

The emitter voltage is attached to the i-f inputs. As it varies, it tends to turn the i-f stages off when it goes positive and on when it goes in a negative direction. Also connected to the emitter is the pnp emitter lead from the agc delay.

118 AGC DELAY CIRCUIT

The agc delay base goes to ground through the agc delay control (Fig. 157). The collector goes to the agc terminal of the tuner. The agc delay reduces the rf amplifier output if and when the signal becomes so strong it tends to overload. The agc delay circuit should not turn on until the i-f agc is at maximum. It's best to adjust the rf only during extremely strong signal conditions. Early agc delay action, which reduces rf amplification prematurely can weaken the signal and bring noise into the picture.

The potentials on the delay are about 2V on the collector and about 5V on the emitter and base. This reverse biases the pnp transistor. As a strong signal comes into the TV, the agc output emitter voltage rises to 6 or 7V. This forward biases the pnp and electrons start flowing from collector to base and emitter. The collector voltage rises above the 2V. The positive voltage increases in proportion to the signal strength applied to the rf amplifier and reduces its conduction. The agc delay

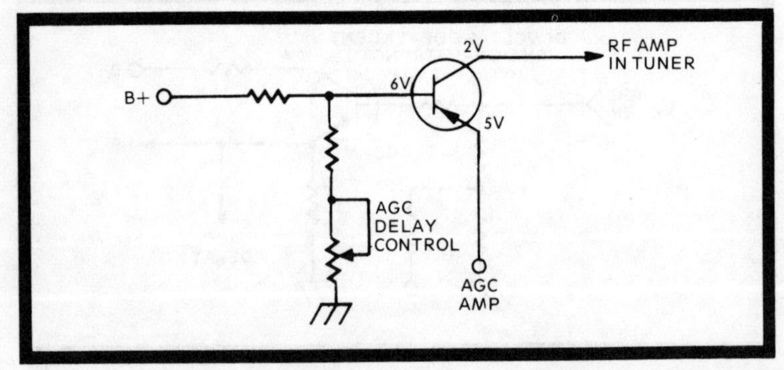

Fig. 157. The agc delay circuit reduces the rf amplifier gain when signal strengths approach overloading.

control is adjusted so that the base voltage is at such a level that the delay will not turn on until the i-f stages have reached a maximum reduction of conduction.

SYNC SEPARATOR CIRCUIT 119

The sync separator is located near the agc circuit. It is an npn transistor with about 10V on the collector and reverse bias on the emitter and base (Fig. 158). The composite video signal is applied to the base through an RC network. The signal causes the bias to go in a forward direction. As the large positive horizontal sync pulse arrives, the base is driven into conduction.

The sync pulse appears in the collector circuit, but the rest of the video is not positive enough to cause conduction. The positive sync signals are inverted 180 degrees in the collector circuit. The sync output is applied to the horizontal and vertical oscillator input circuits.

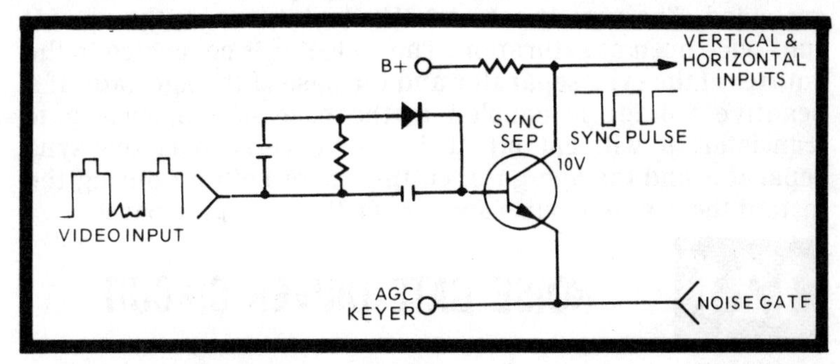

Fig. 158. The sync separator clips sync pulses off the pedestal and inverts the pulses 180 degrees.

NOISE GATE CIRCUIT 120

A big problem in both agc and sync separation circuits is noise. Most noise is of an AM nature and can get into the video signal. It is then detected and applied to all the circuits that need video to produce designed outputs. Special noise immunity circuits are needed in the agc and sync stages; otherwise, the TV picture becomes unstable and erratic.

The agc and sync circuits do two different jobs, but they can be protected from noise by a common noise immunity

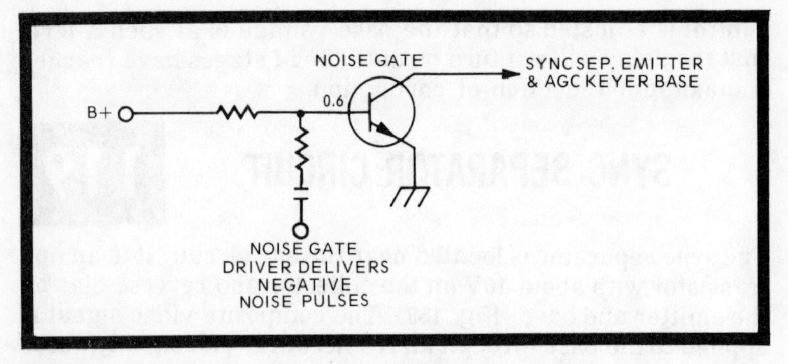

Fig. 159. A noise gate circuit shuts down the conduction of the agc and sync stages during the presence of noise pulses.

circuit. The noise circuits cut off the agc and sync circuits during the intervals noise appears. Most of the time noise appears in the video. Should noise occur during the horizontal sync pulse, the noise circuits will shut everything down during those one or two pulses, which is quite satisfactory.

The noise gate (Fig. 159) is an npn with the emitter grounded. The base is set at 0.6V which turns on the npn. It runs wide open at saturation. The collector is connected to the emitter of the sync separator and the base of the agc gate. If a negative voltage is coupled to the base of the noise gate transistor, it will cut off. This will reverse bias the sync separator and the agc gate, cutting them both off during the instant the negative pulse appears on the noise gate base.

121 NOISE GATE DRIVER CIRCUIT

The noise gate driver circuit has the job of getting a negative pulse to the noise gate base when noise appears during the horizontal sync pulse interval. The noise gate driver, also an npn, has B-plus of about 20V on its collector (Fig. 160). The base voltage is applied through the noise gate control. The control is adjusted to just cut off the driver.

Negative-going video is applied to the emitter of the driver. The bias is set to keep the npn cut off at levels as high as horizontal sync pulse. Should a noise pulse appear that is even more negative than the sync pedestal, only then will the transistor conduct. As it conducts, the amplified negative-going noise (no phase change from emitter to collector) is coupled capacitively into the base of the noise gate transistor. The negative pulse cuts off the noise gate, raising its collector voltage, which causes reverse bias and cutoff in the sync separation and agc gate circuits. Noise cannot interfere with the operation of two circuits.

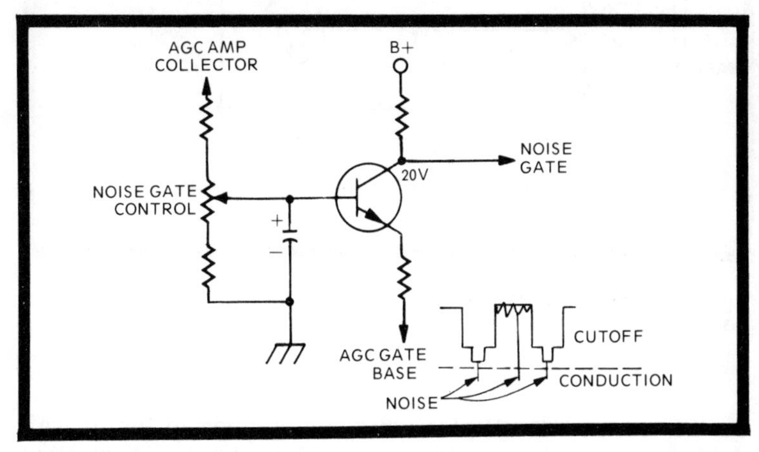

Fig. 160. A noise gate driver clips noise pulses off the sync tips and sends them to the noise gate.

TRANSFORMERLESS AUDIO OUTPUT 122

In large stereo output circuits (those above 3 or 4 watts), the operating efficiency of the circuit is one of the most important considerations. The power that is dissipated in proportion to the amount of audio must be kept low.

Class AB amplifiers provide the best operating efficiency, but such amplifiers need an expensive audio output transformer. In order to transfer a flat response from the lowest lows to the highest highs, the expense of the transformer and its size almost eliminates the feasibility of a commercial system. One way to get rid of the output transformer is to match the audio output into the speaker impedance directly from the transistor. This is done in a 2-transistor simple circuit.

An ordinary pnp transistor is attached to a power npn transistor. The collector of the pnp connects to the base of the npn (Fig. 161). The emitter of the pnp connects to the collector of the npn. In the npn a base-leak resistor connects from base to emitter. The 2-transistor setup becomes, for all intents and purposes, one transistor with four pn junctions. The overall emitter of the configuration is the emitter of the pnp. The overall collector is the emitter of the npn.

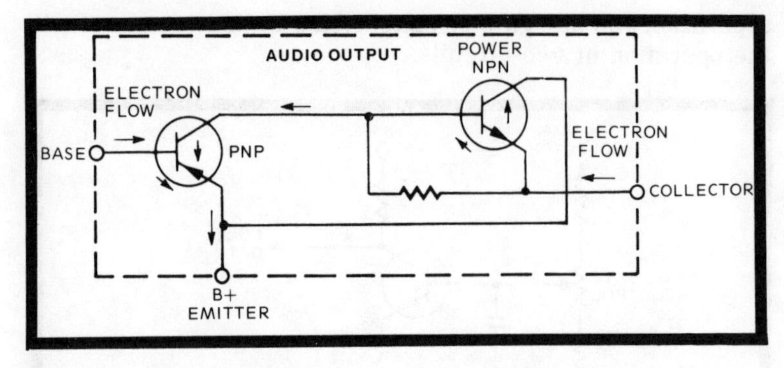

Fig. 161. Two transistors connected as shown act like one power pnp and match a speaker impedance.

When a positive B-plus voltage is applied to the overall emitter, and the base is biased in a forward direction, electrons flow from the base to the emitter. Electrons flow from the overall collector, through the npn to the emitter. Also electrons flow from the overall collector to the npn base, to the pnp collector, and to the base and emitter of the pnp. The configuration acts like a high power pnp transistor.

There is no problem with current gain in this overall emitter follower; and in power output circuits, it is current gain for power, rather than voltage gain, that is desired.

The emitter follower can be matched easily into any impedance desired. Typically it is 8 or 16 ohms.

123 HALF-WAVE ANTENNA CIRCUIT

The commonest type of receiving antenna is the half wave antenna. The receiving antenna can be thought of as the secondary of a gigantic transformer. The primary is the transmitting antenna that puts an electromagnetic field into space as current moves back and forth in it at the rf. The receiving antenna intercepts the transmitting field and in turn has a current developed in it.

If the antenna is cut to be half the wavelength of the signal, it resonates at the signal frequency and develops a maximum current that can be measured in microamps. As the electrons move back and forth in the half wave, the electron movement is greatest at the center and minimum at the ends. Since electron flow or current is greatest at the center of the an-

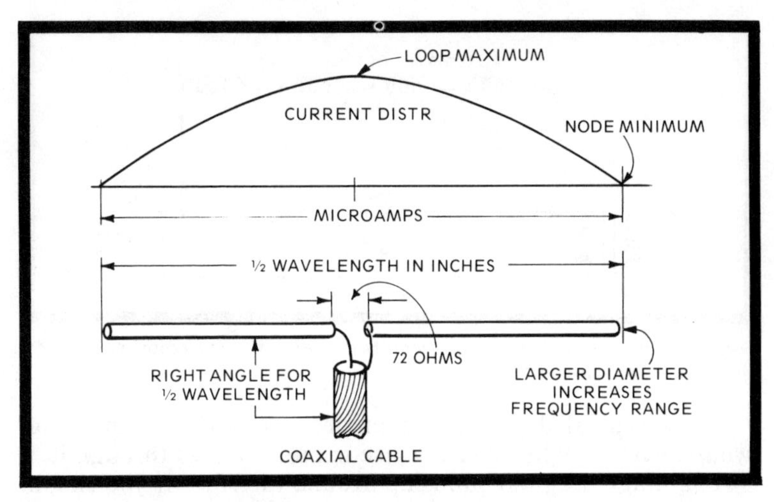

Fig. 162. A half-wave antenna circuit develops maximum current at its center with an impedance of about 72 ohms.

tenna, the resistance is lowest at the center. The resistance is calculated to be about 72 ohms. As the current travels away from the center, the resistance increases. For the best impedance match a 72-ohm coaxial cable can be attached to the center of the half-wave antenna, the point of maximum electron movement.

If the diameter of the antenna is increased, the amount of capacity in the antenna is increased. As C increases, the LC ratio decreases. The increased capacitance causes the resonance to decrease. This permits the antenna to operate over a broader frequency range. For instance, a black-and-white TV antenna can have smaller rods than a color antenna, since the color antenna must receive a broader bandpass.

AUTOMATIC DEGAUSSING CIRCUIT 124

Take a small pocket magnet and hold it near a TV display. At the point where you place the magnet the picture distorts. If a portion of the picture tube or its hardware should become magnetized, it will distort the picture in the same way. It's annoying in a black-and-white picture and it ruins a color picture.

The old metal shell black-and-white tubes were subject to this trouble. To remove the fixed distortion, a demagnetizer had to be used. Since magnetic fields are measured in gauss, demagnetizers are aptly named degaussers.

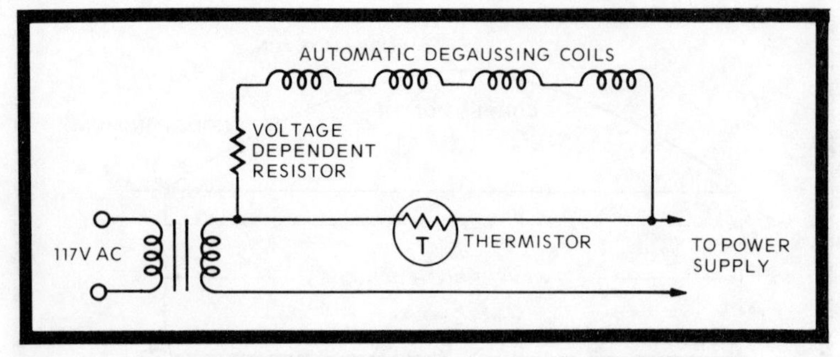

Fig. 163. Automatic degaussing circuits demagnetize the color crt every time the TV is turned on.

The typical degausser is an iron core ring with a coil wound over it. When 60 Hz house current passes through it, a strong magnetic field builds up around the ring. If you should pass the magnetic field in a circle around a magnetized crt area, the area of demagnetization will disappear.

It became a valuable accessory to install ADG (automatic degaussing) to color TVs. The circuit consists of several degaussing coils formed around the picture tube shell, a thermistor, and a voltage-dependent resistor (VDR). The termistor (a negative temperature coefficient resistor) is attached in series between the secondary of the power transformer and the rectifier. The VDR and coils are connected across the thermistor (Fig. 163).

The thermistor resistance decreases as current passes through it and causes it to heat up. The thermistor typically starts off at a couple hundred ohms cold and goes to near zero ohms hot.

The VDR resistance is near zero ohms while the thermistor is cold and then increases in resistance up to 10K as the thermistor gets hot and approaches zero ohms. The VDR goes up in resistance as the thermistor drops in resistance. Therefore, as power is turned on, the thermistor and its parallel circuit (the VDR and coils) present two paths to the electrons. Since the thermistor is a high resistance and the VDR coils are a relatively low resistance, most of the electrons flow through the coils, causing a large magnetic field to be induced. This automatically degausses the crt and its environs.

As the thermistor heats up and decreases in resistance, more electrons flow through it and less through the VDR coil leg. The VDR increases in resistance, causing even less current through it and the coil. When the thermistor is hot and near zero ohms, the VDR resistance is near 10K. Practically all of the electrons now go through the thermistor. For all intents, the degausser is off and the TV plays on.

Adg coils are weak and get rid of small magnetic fields that develop. Any strong magnetic fields must still be eliminated with the large hand-held ring degausser.

AUTOMATIC BRIGHTNESS CIRCUIT 125

An optional feature in some TVs is an automatic brightness circuit. Actually, the circuit adjusts brightness and contrast by varying the emitter resistance in the video output circuit.

The video is coupled to the crt cathode so that the sync tips are positive and will blank out the picture. The video output transistor is dc-coupled to the crt cathode. Any changes in video gain changes the dc bias between the cathode and control grid of the crt, thus changing both the brightness and contrast.

The contrast control is in the base of the output (Fig. 164). The brightness control is in the cathode of the crt. The abc control is in the screen grid of the crt. All three controls vary the crt brightness by changing the crt bias. The contrast control, in addition, varies the contrast, too.

The key to the abc circuit is a light-dependent resistor (LDR) in the output emitter circuit in a parallel with another emitter resistor. The LDR is not bypassed so it can have full

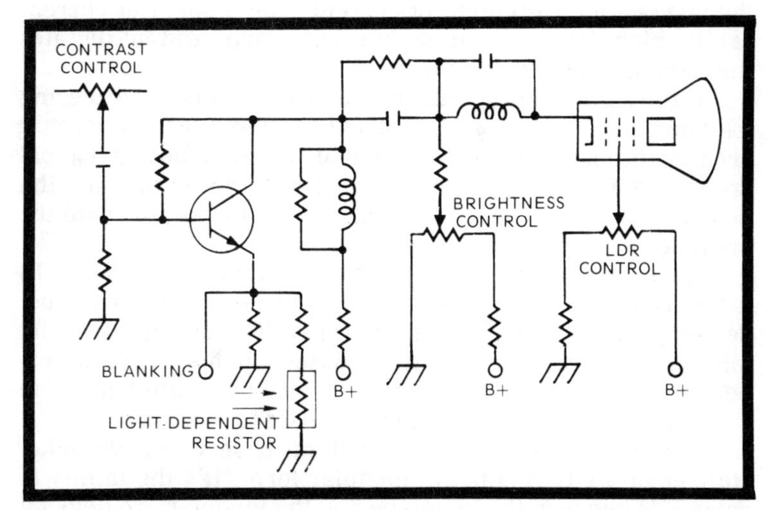

Fig. 164. Automatic brightness circuits are operated by the amount of light that shines on the LDR.

effect. The LDR is mounted on the bottom of the crt and is exposed to the ambient lighting through a hole in the cabinet.

If the room is bright, the LDR resistance drops. The lower resistance causes a drop in the emitter voltage. The npn conducts more, increasing the gain and the picture brightness and contrast. If the room is dark the LDR increases in resistance. This raises the emitter voltage, reducing its gain. The crt cathode goes a bit more positive, increasing the crt bias and the display loses brightness and contrast. In the TV viewing room, as the afternoon merges into evening, the TV picture brightness and contrast change accordingly.

FLOW CHARTS

The servicer must understand the circuit he is about to troubleshoot. He must "see" the electrons moving from place to place. Once he has an idea of what's happening in the devices, he can pick out circuit areas where all is normal or not normal. At that time, his mind should operate like a computer. He makes an indicated test and from the test results comes to a conclusion. If the test is a scope picture, the scope shows a normal waveform, a distorted waveform, or no waveform. Each situation indicates the next move. If the test is a dc voltage, the correct or incorrect voltage will be present. Each voltage indication indicates the next move. Or perhaps the test is a measurement of current. The amount of current can be high, low, normal, or missing. Each reading dictates the next logical service move.

If the servicer proceeds in a logical manner by a piecing together and nonclues, he quickly arrives at the circuit area containing the trouble. Should he not follow a logical procedure, but in scattershot fashion just tries to pick out the defect, only sheer chance will enable him to stumble onto the breakdown.

Each basic trouble symptom has its own flow chart. In many cases only the schematic is available and you must devise your flow chart. The experienced servicer naturally follows a logical chart pattern based on his years of experience. With a little practice the novice can train himself to think in flow-chart fashion. All he needs is to be shown how.

One of the toughest electronic items to service is the solid-state color TV that is not in modular form. It's the family of color TVs between the tube-type and the modular. To help TV servicemen repair these complex, miniaturized pieces, manufacturers have compiled flow charts, in addition to the schematic and ordinary service notes.

With the idea in mind that, if you can troubleshoot these complex color TV circuits with the aid of a flow chart, you can then dream up your own flow chart for the less complicated types of gear, following are the charts for a GE 19-inch color receiver. These charts were made by drawing sixteen blocks on a piece of paper.

The trouble symptom is recorded at the top. A test is made. The tests are voltage, current, waveform, and visual. A conclusion is drawn from the results of the test. The conclusions are followed, more tests are made, and the end result is the defective circuit amd component. A careful analysis of how the GE people did it should enable a lot of servicers to produce their own flow charts, either on paper or in the mind's eye, to speed the repair of electronic equipment.

APPENDIX I

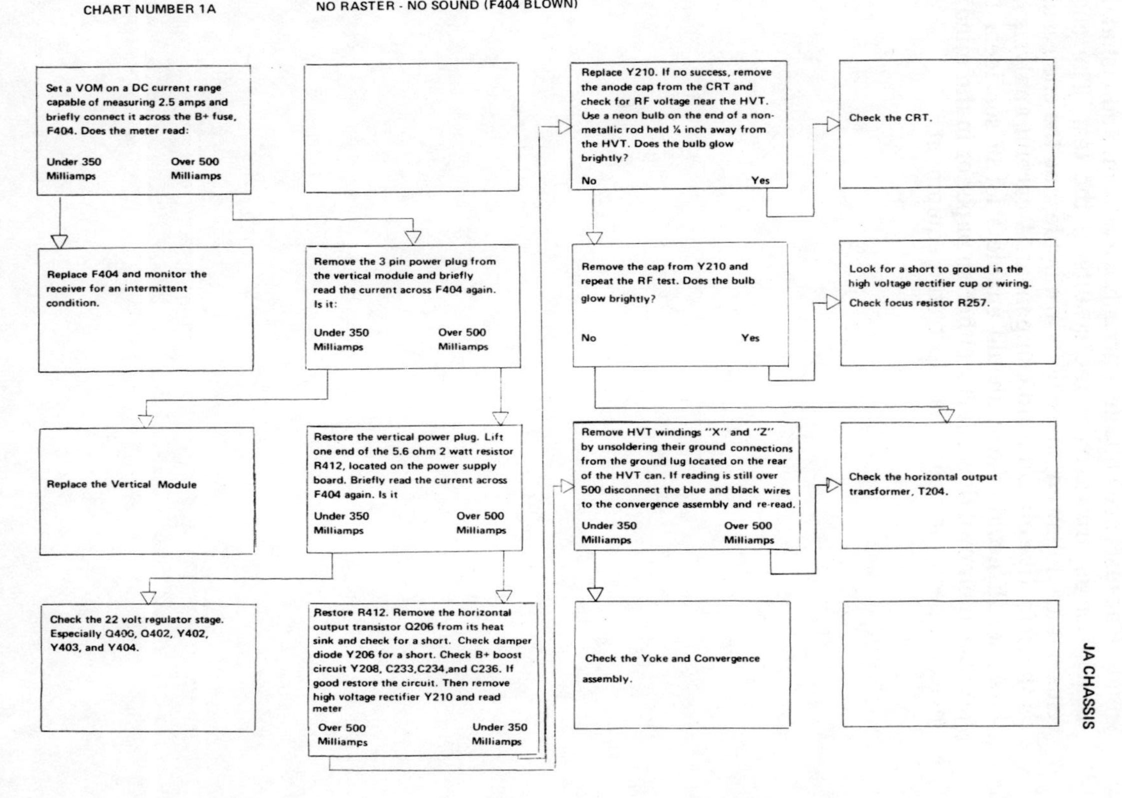

JA CHASSIS
CHART NUMBER 1A
NO RASTER - NO SOUND (F404 BLOWN)
Set a VOM on a DC current range capable of measuring 2.5 amps and briefly connect it across the B+ fuse, F404. Does the meter read:
Under 350 Milliamps
Over 500 Milliamps
Replace F404 and monitor the receiver for an intermittent condition.
Remove the 3 pin power plug from the vertical module and briefly read the current across F404 again. Is it:
Under 350 Milliamps
Over 500 Milliamps
Replace the Vertical Module
Restore the vertical power plug. Lift one end of the 5.6 ohm 2 watt resistor R412, located on the power supply board. Briefly read the current across F404 again. Is it
Under 350 Milliamps
Over 500 Milliamps
Check the 22 volt regulator stage. Especially Q40G, Q402, Y402, Y403, and Y404.
Restore R412. Remove the horizontal output transistor Q206 from its heat sink and check for a short. Check damper diode Y206 for a short. Check B+ boost circuit Y208, C233,C234,and C236. If good restore the circuit. Then remove high voltage rectifier Y210 and read meter
Over 500 Milliamps
Under 350 Milliamps
Replace Y210. If no success, remove the anode cap from the CRT and check for RF voltage near the HVT. Use a neon bulb on the end of a non-metallic rod held ¼ inch away from the HVT. Does the bulb glow brightly?
No
Yes
Check the CRT.
Remove the cap from Y210 and repeat the RF test. Does the bulb glow brightly?
No
Yes
Look for a short to ground in the high voltage rectifier cup or wiring. Check focus resistor R257.
Remove HVT windings "X" and "Z" by unsoldering their ground connections from the ground lug located on the rear of the HVT can. If reading is still over 500 disconnect the blue and black wires to the convergence assembly and re-read.
Under 350 Milliamps
Over 500 Milliamps
Check the horizontal output transformer, T204.
Check the Yoke and Convergence assembly.

NO RASTER — NO SOUND (F404 INTACT)

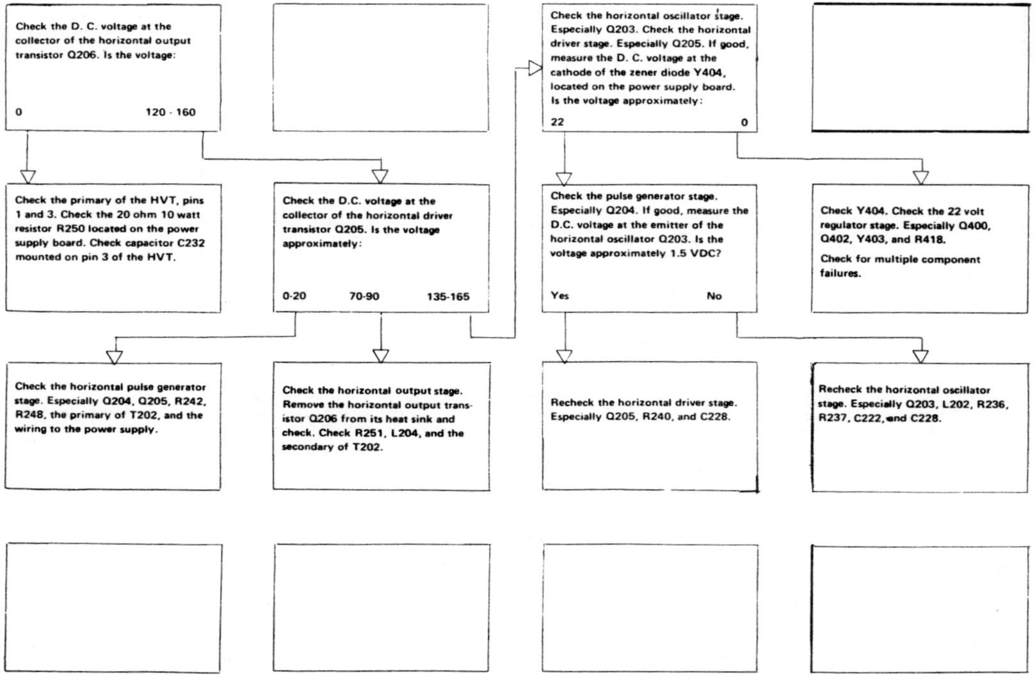

Check the D. C. voltage at the collector of the horizontal output transistor Q206. Is the voltage:

0 | 120 - 160

Check the primary of the HVT, pins 1 and 3. Check the 20 ohm 10 watt resistor R250 located on the power supply board. Check capacitor C232 mounted on pin 3 of the HVT.

Check the D.C. voltage at the collector of the horizontal driver transistor Q205. Is the voltage approximately:

0-20 | 70-90 | 135-165

Check the horizontal pulse generator stage. Especially Q204, Q205, R242, R248, the primary of T202, and the wiring to the power supply.

Check the horizontal output stage. Remove the horizontal output transistor Q206 from its heat sink and check. Check R251, L204, and the secondary of T202.

Check the horizontal oscillator stage. Especially Q203. Check the horizontal driver stage. Especially Q205. If good, measure the D. C. voltage at the cathode of the zener diode Y404, located on the power supply board. Is the voltage approximately:

22 | 0

Check the pulse generator stage. Especially Q204. If good, measure the D.C. voltage at the emitter of the horizontal oscillator Q203. Is the voltage approximately 1.5 VDC?

Yes | No

Check Y404. Check the 22 volt regulator stage. Especially Q400, Q402, Y403, and R418.

Check for multiple component failures.

Recheck the horizontal driver stage. Especially Q205, R240, and C228.

Recheck the horizontal oscillator stage. Especially Q203, L202, R236, R237, C222, and C228.

NO RASTER

CHART NUMBER 2

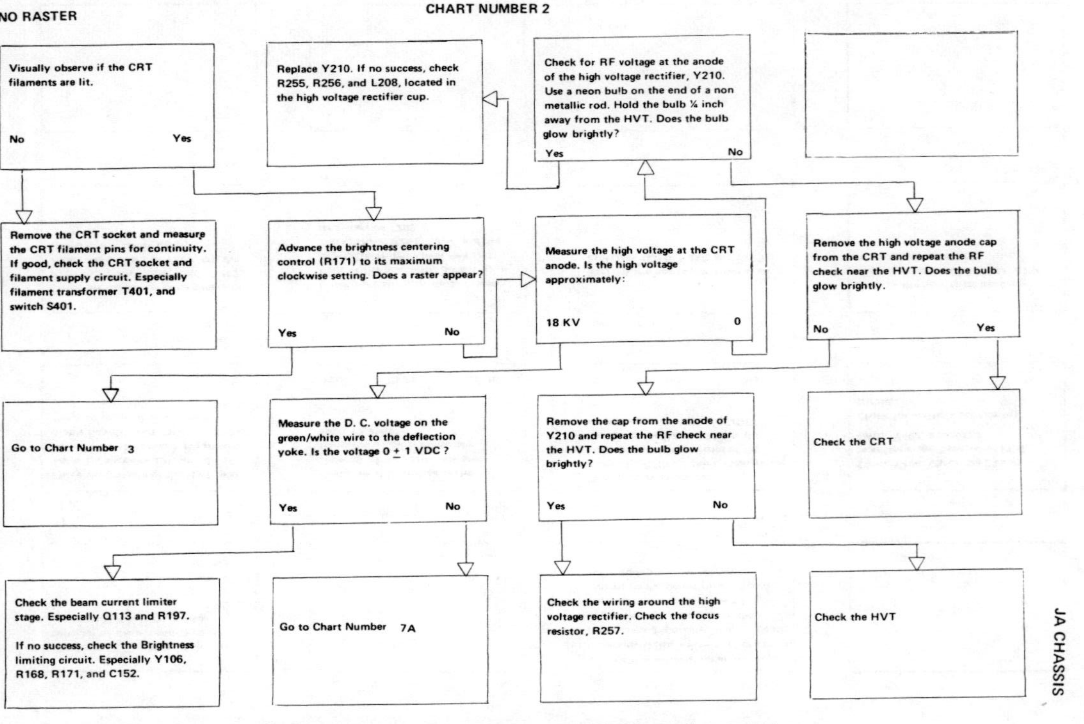

*NOTE: After the receiver has been repaired, reset Brightness Centering (R171) by the adjustment procedure.

NO VIDEO

CHART NUMBER 3

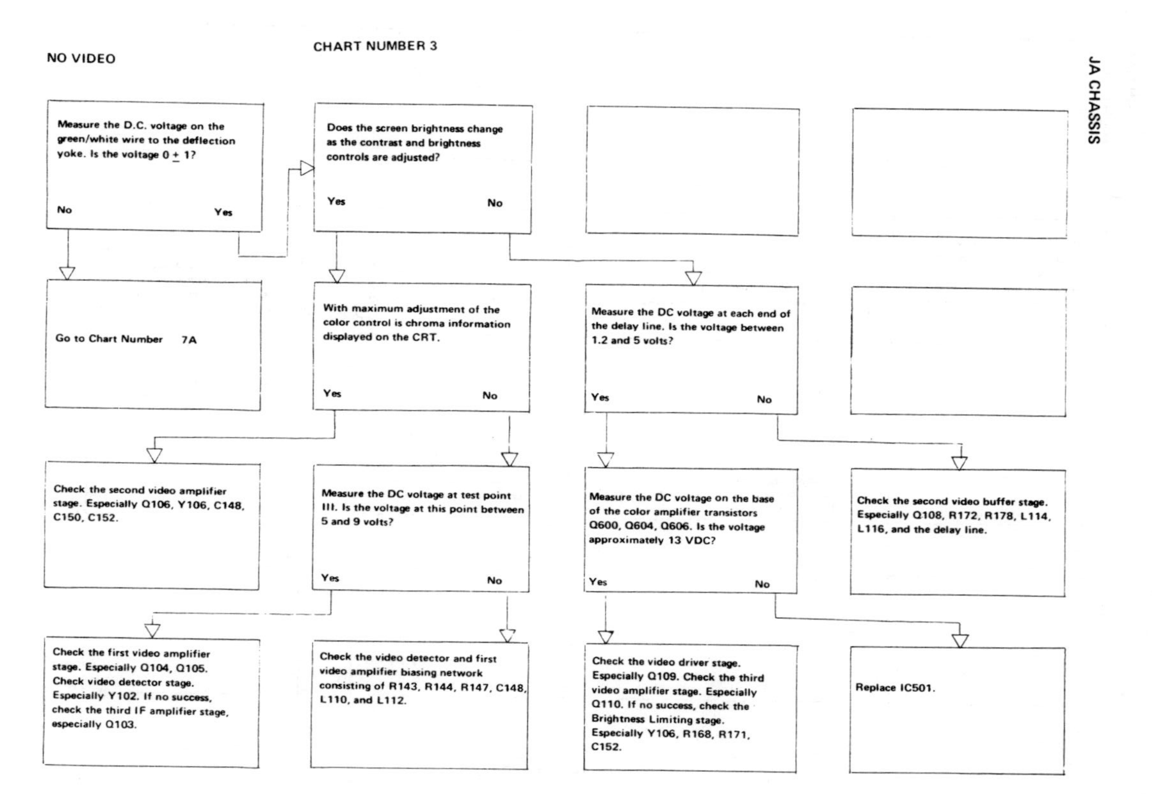

CHART NUMBER 4

NO VIDEO — NO AUDIO

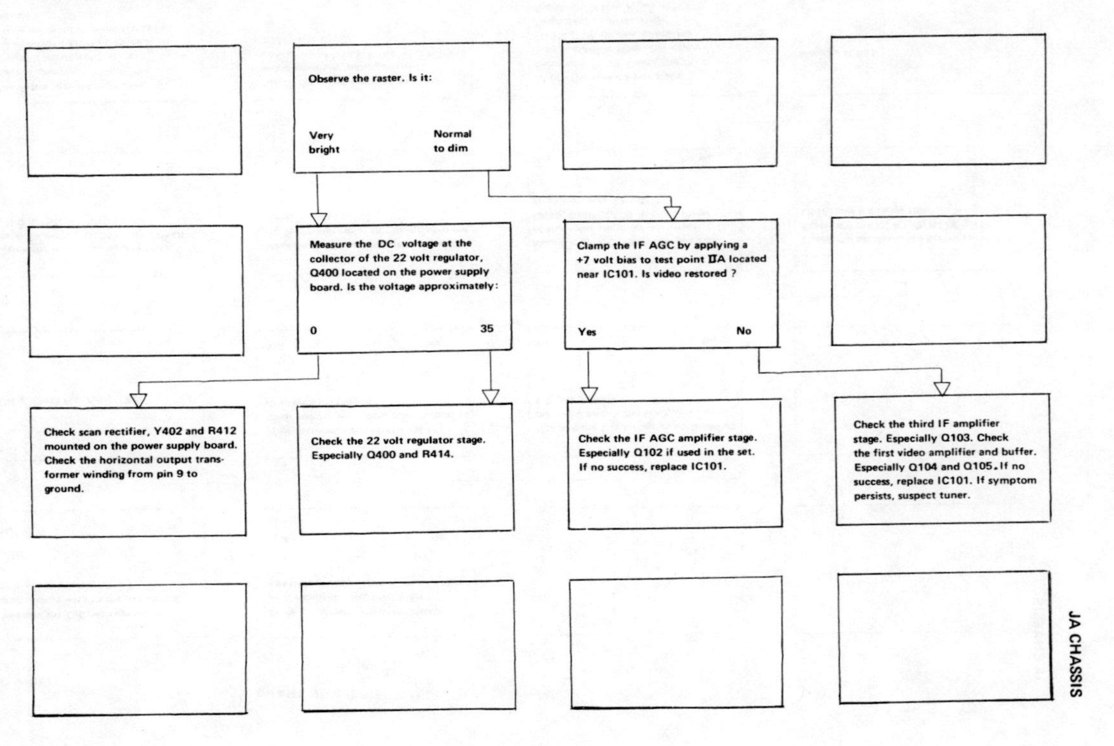

SNOWY, DISTORTED, OR OVERLOADED VIDEO

CHART NUMBER 5

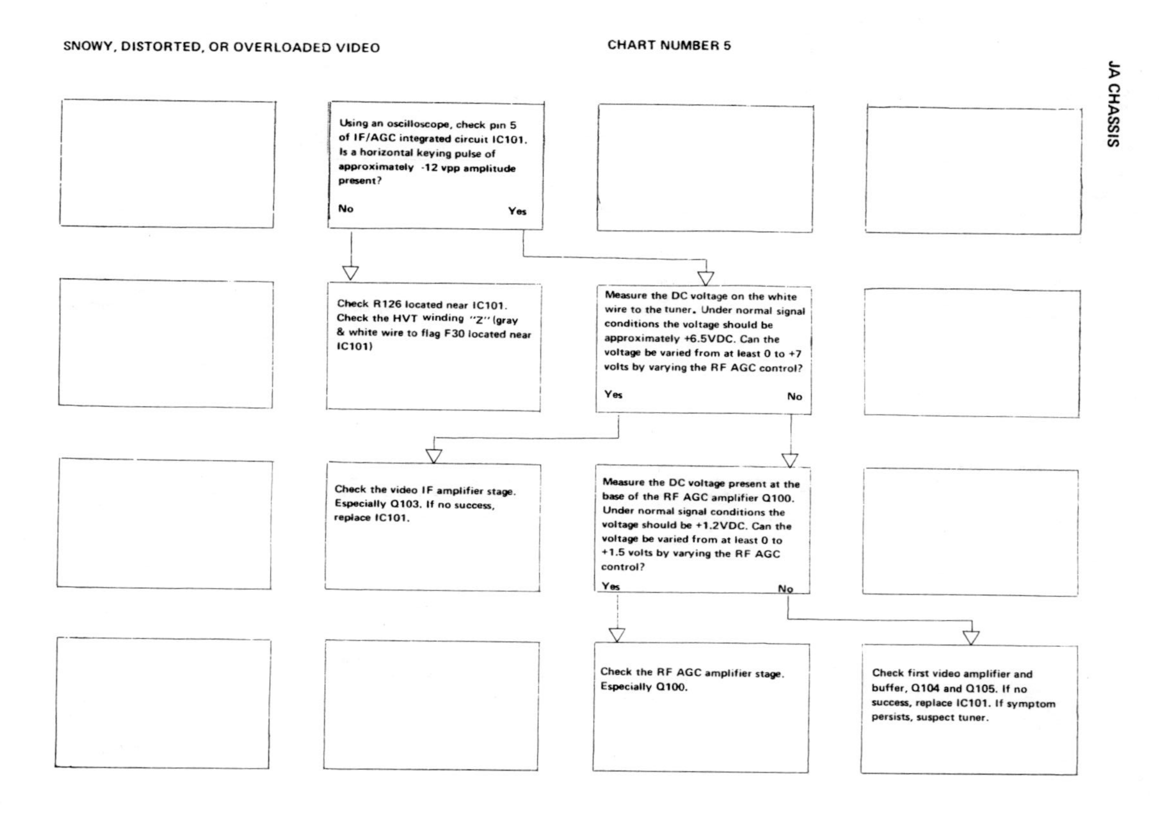

SYNC PROBLEMS

CHART NUMBER 6

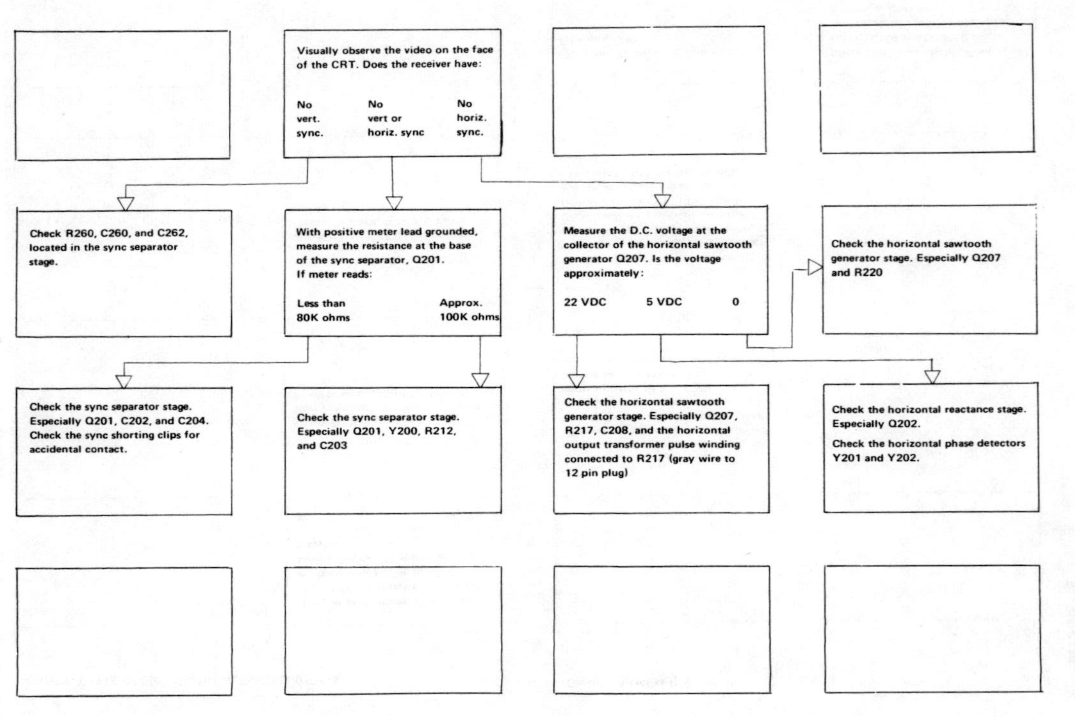

JA CHASSIS

VERTICAL MODULE CHECK

CHART NUMBER 7A

CHART NUMBER 8

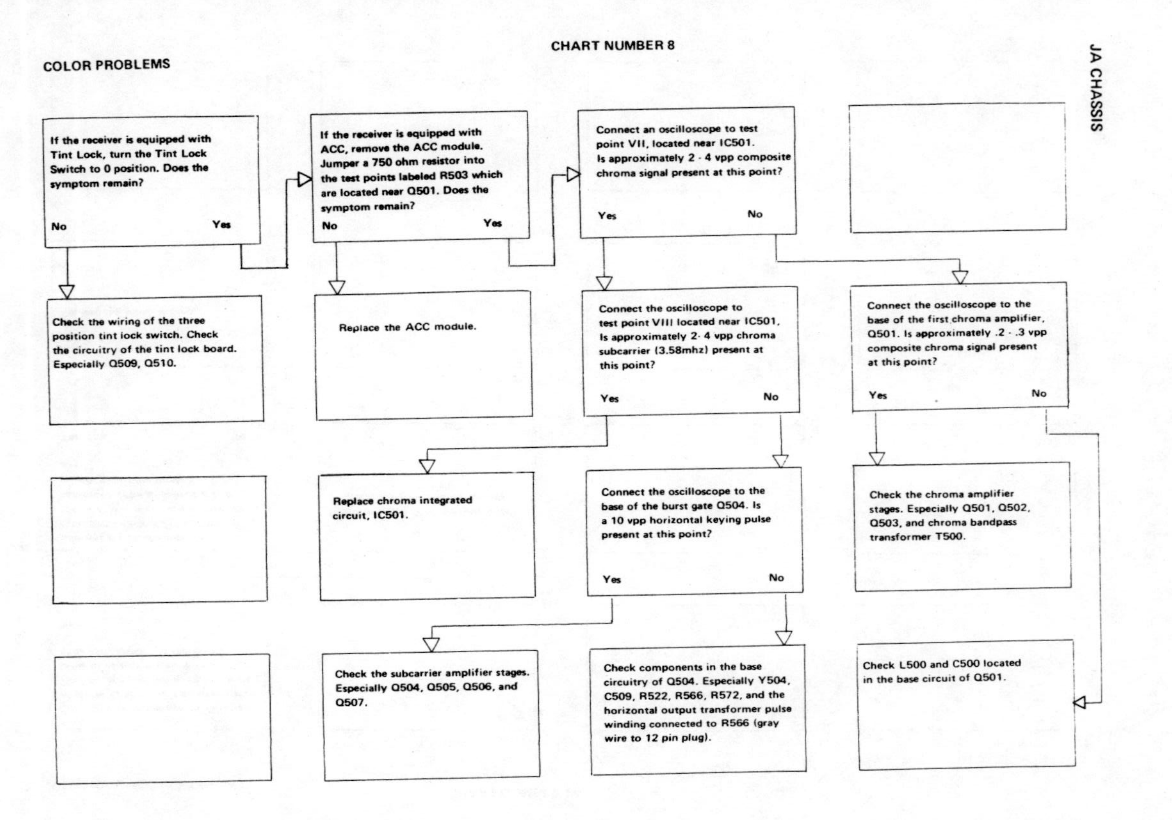

GRAY SCALE PROBLEMS

CHART NUMBER 9

Switch CRT grid 1 wire of affected color with one of the other CRT grid 1 wires. Does the symptom change

No Yes

Switch CRT grid 2 wire of affected color with one of the other CRT grid 2 wires. Does the symptom change?

No Yes

Restore the CRT grid 1 wires to their original test points. Switch the color amplifier transistor of the affected color with one of the other color amplifier transistors. Does the symptom change?

No Yes

Switch CRT cathode wire of affected color with one of the other CRT cathode wires. Does the symptom change?

No Yes

Restore the CRT grid 2 wires to their original test points. Check the grid 2 circuitry of the affected color. Especially screen controls R640, R642, R644

Check the circuitry of the affected color amplifier. Especially check diodes Y600, Y602, Y604, Y606, Y608, and Y610. If no success, replace IC501.

Replace the defective color amplifier transistor. (Q600, Q604, or Q606).

Check CRT socket. If no success, suspect CRT.

Restore the CRT cathode wires to their original test points. Check the CRT cathode circuitry of the affected color. Especially drive controls R194, R195, R196.

CHART NUMBER 10

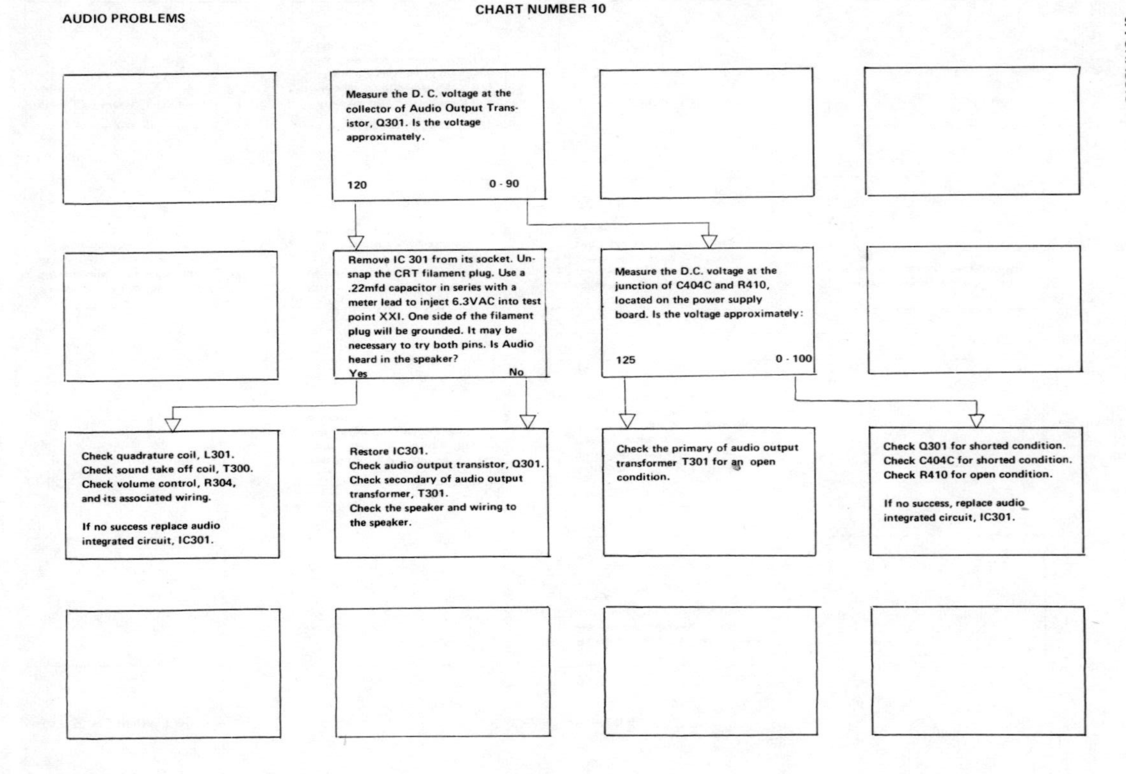

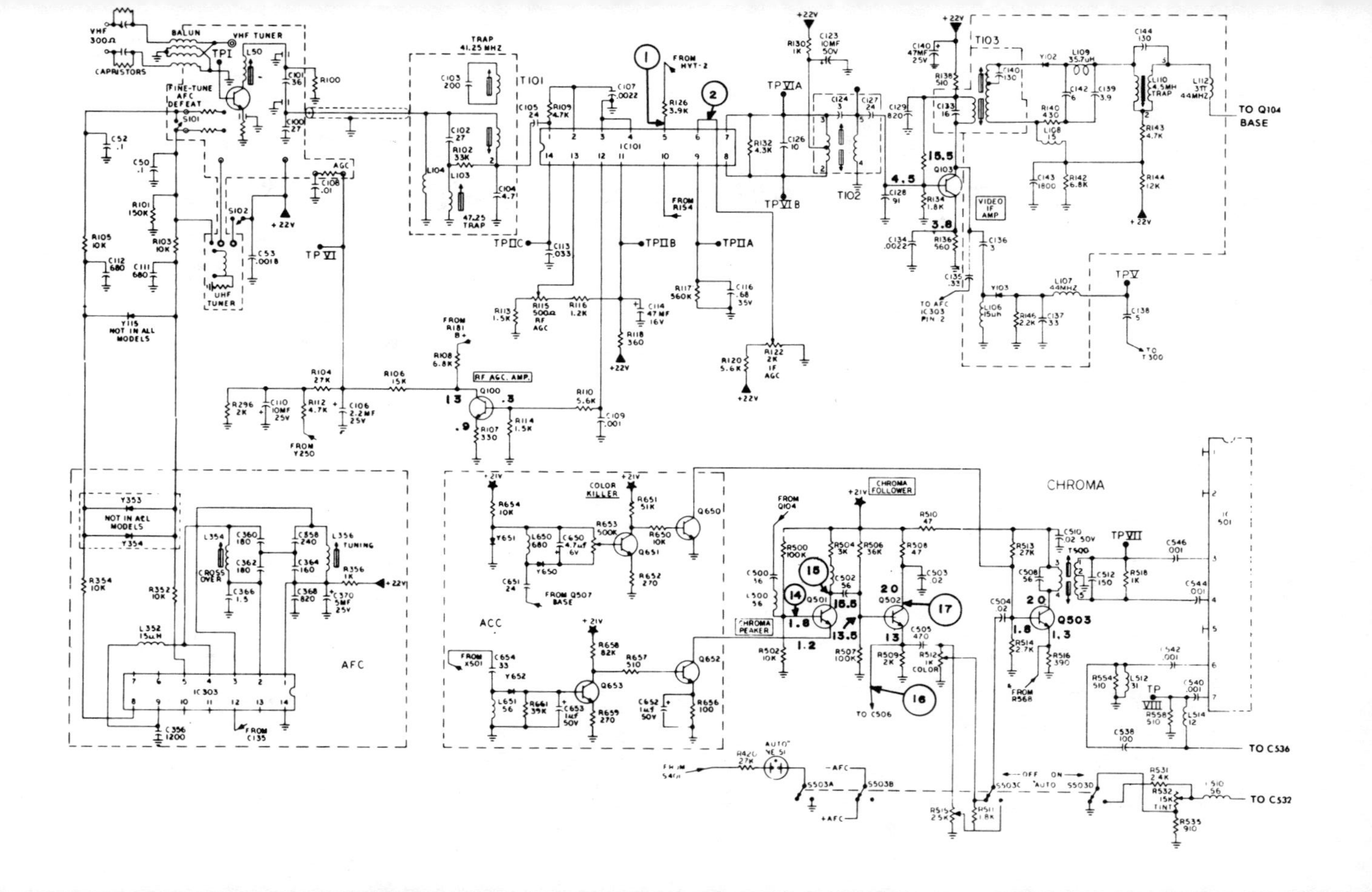
VHF 300Ω
CAPRISTORS
BALUN
VHF TUNER
TPI
FINE-TUNE AFC DEFEAT
S101
UHF TUNER
S102
+22V
AGC
TP VI
Y115 NOT IN ALL MODELS
TRAP 41.25 MHZ
47.25 TRAP
T101
IC101
FROM HVT-2
FROM R154
TP II C
TP II B
TP II A
R115 500Ω RF AGC
R122 2K IF AGC
TP VI A
TP VI B
T102
T103
Q103
VIDEO IF AMP
TO AFC IC303 PIN 2
L110 4.5MH TRAP
L112 3T 44MHZ
TO Q104 BASE
L107 44MHZ
TP V
TO T300
FROM R181 B+
RF AGC. AMP.
Q100
FROM Y250
Y353
NOT IN ALL MODELS
Y354
CROSS OVER
L356 TUNING
IC303
FROM C135
AFC
COLOR KILLER
Q650
Q651
FROM Q507 BASE
FROM X501
Q652
Q653
ACC
CHROMA FOLLOWER
FROM Q104
CHROMA PEAKER
Q501
Q502
Q503
TO C506
FROM R568
CHROMA
TP VII
T500
IC 501
TP VIII
TO C536
TO C532
R512 1K COLOR
R532 15K TINT
AUTO NE 51
FROM S401
S503A
S503B
S503C
S503D
-AFC
+AFC
OFF ON
AUTO

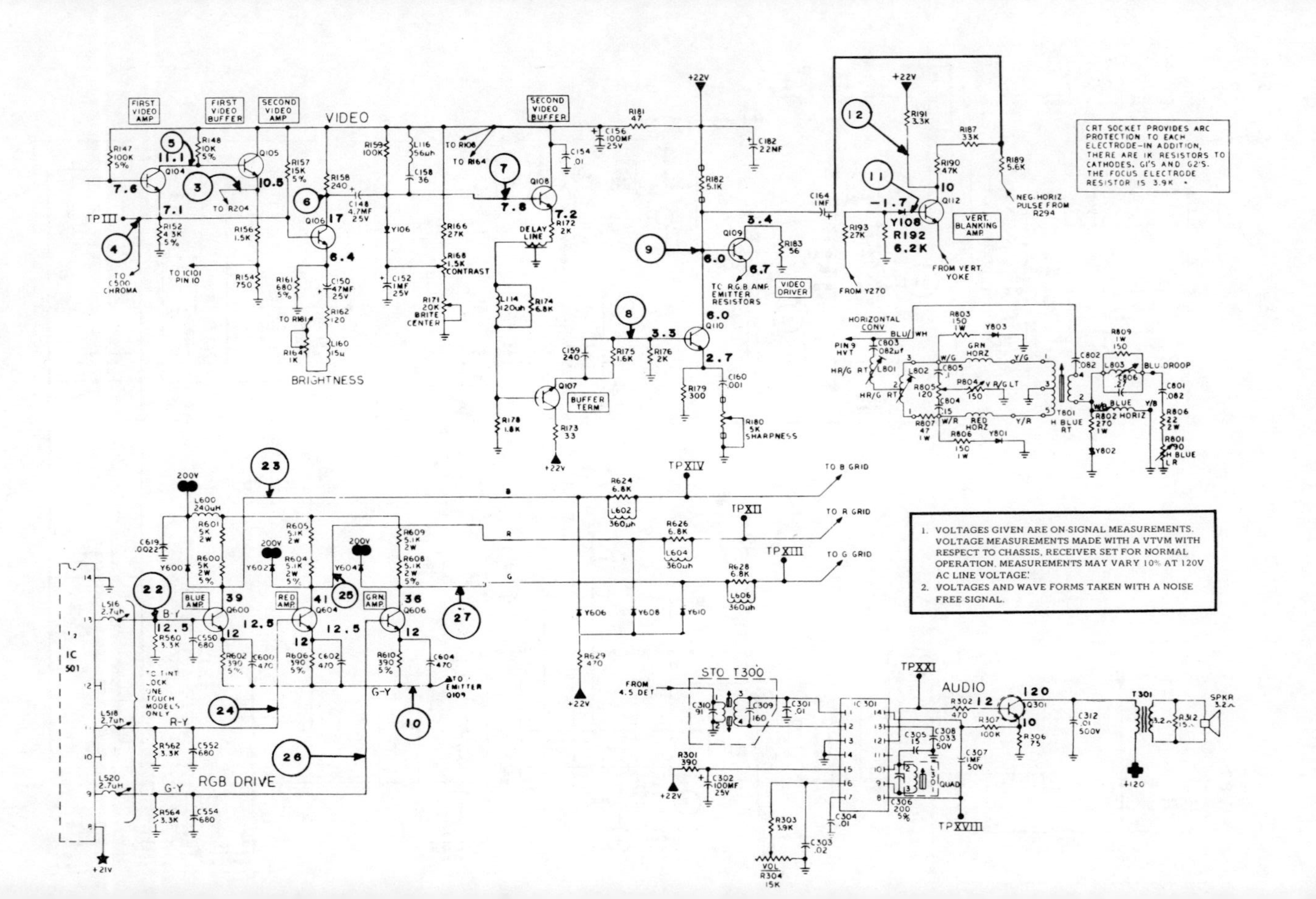
CRT SOCKET PROVIDES ARC PROTECTION TO EACH ELECTRODE—IN ADDITION, THERE ARE 1K RESISTORS TO CATHODES, G1'S AND G2'S. THE FOCUS ELECTRODE RESISTOR IS 3.9K.
1. VOLTAGES GIVEN ARE ON-SIGNAL MEASUREMENTS. VOLTAGE MEASUREMENTS MADE WITH A VTVM WITH RESPECT TO CHASSIS, RECEIVER SET FOR NORMAL OPERATION. MEASUREMENTS MAY VARY 10% AT 120V AC LINE VOLTAGE.
2. VOLTAGES AND WAVE FORMS TAKEN WITH A NOISE FREE SIGNAL.
FIRST VIDEO AMP
FIRST VIDEO BUFFER
SECOND VIDEO AMP
SECOND VIDEO BUFFER
VIDEO
BRIGHTNESS
CONTRAST
DELAY LINE
BUFFER TERM
VIDEO DRIVER
SHARPNESS
VERT. BLANKING AMP
NEG. HORIZ PULSE FROM R294
FROM VERT. YOKE
HORIZONTAL CONV
BLU DROOP
TO B GRID
TO R GRID
TO G GRID
BLUE AMP
RED AMP
GRN AMP
RGB DRIVE
STO T300
FROM 4.5 DET
AUDIO
SPKR
IC 501
IC 301

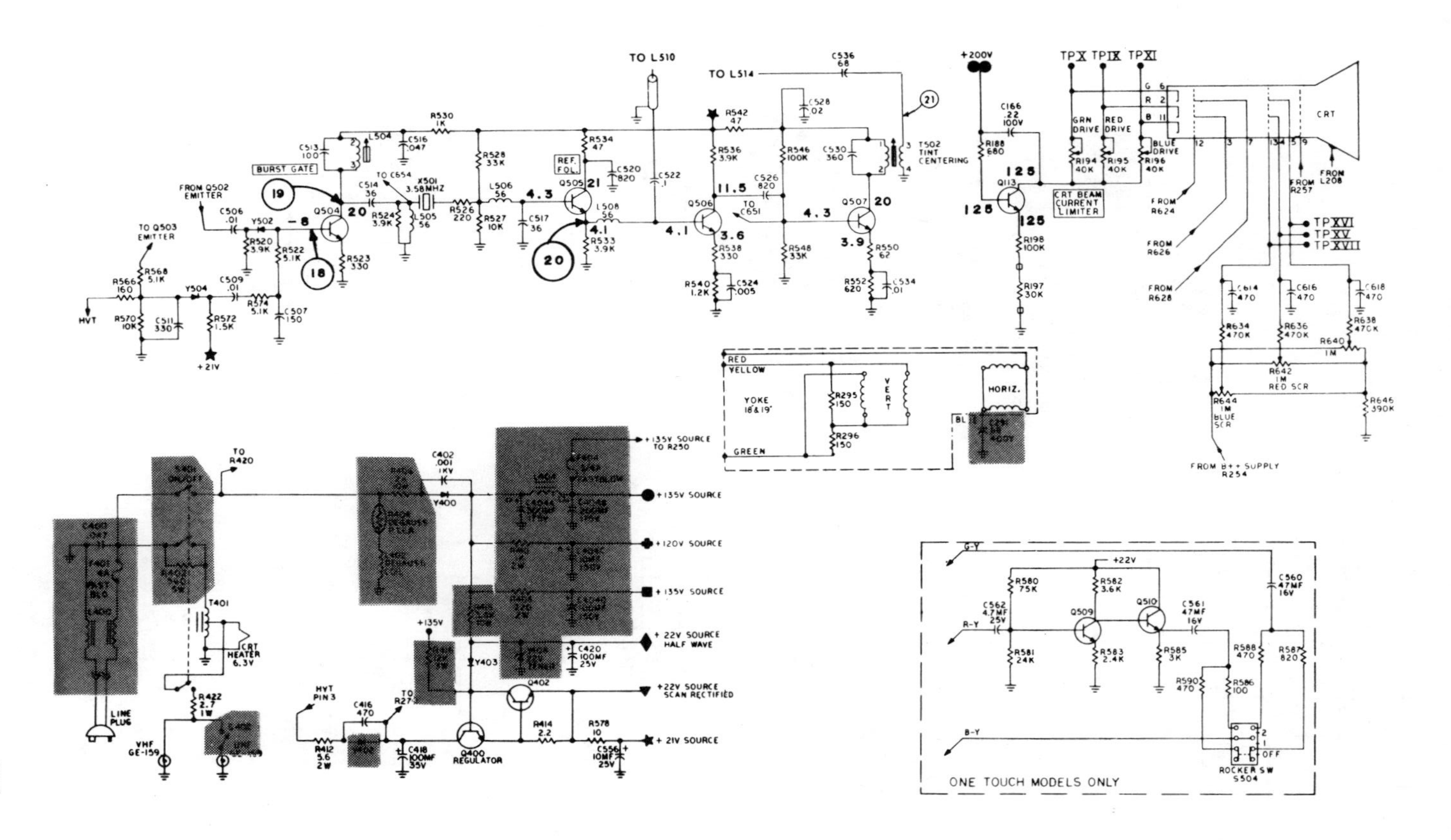
TO L510
TO L514
BURST GATE
FROM Q502 EMITTER
TO Q503 EMITTER
HVT
+21V
TO C654
X501 3.58MHZ
REF. FOL.
TO C651
T502 TINT CENTERING
+200V
CRT BEAM CURRENT LIMITER
GRN DRIVE
RED DRIVE
BLUE DRIVE
CRT
FROM L208
FROM R257
FROM R624
FROM R626
FROM R628
RED SCR
BLUE SCR
FROM B++ SUPPLY R254
RED
YELLOW
YOKE 18&19"
GREEN
VERT
HORIZ.
TO R420
+135V SOURCE TO R250
+135V SOURCE
+120V SOURCE
+135V SOURCE
+22V SOURCE HALF WAVE
+22V SOURCE SCAN RECTIFIED
+21V SOURCE
+135V
HVT PIN 3
Q400 REGULATOR
CRT HEATER 6.3V
LINE PLUG
VHF GE-159
+22V
G-Y
R-Y
B-Y
ROCKER SW S504
OFF
ONE TOUCH MODELS ONLY

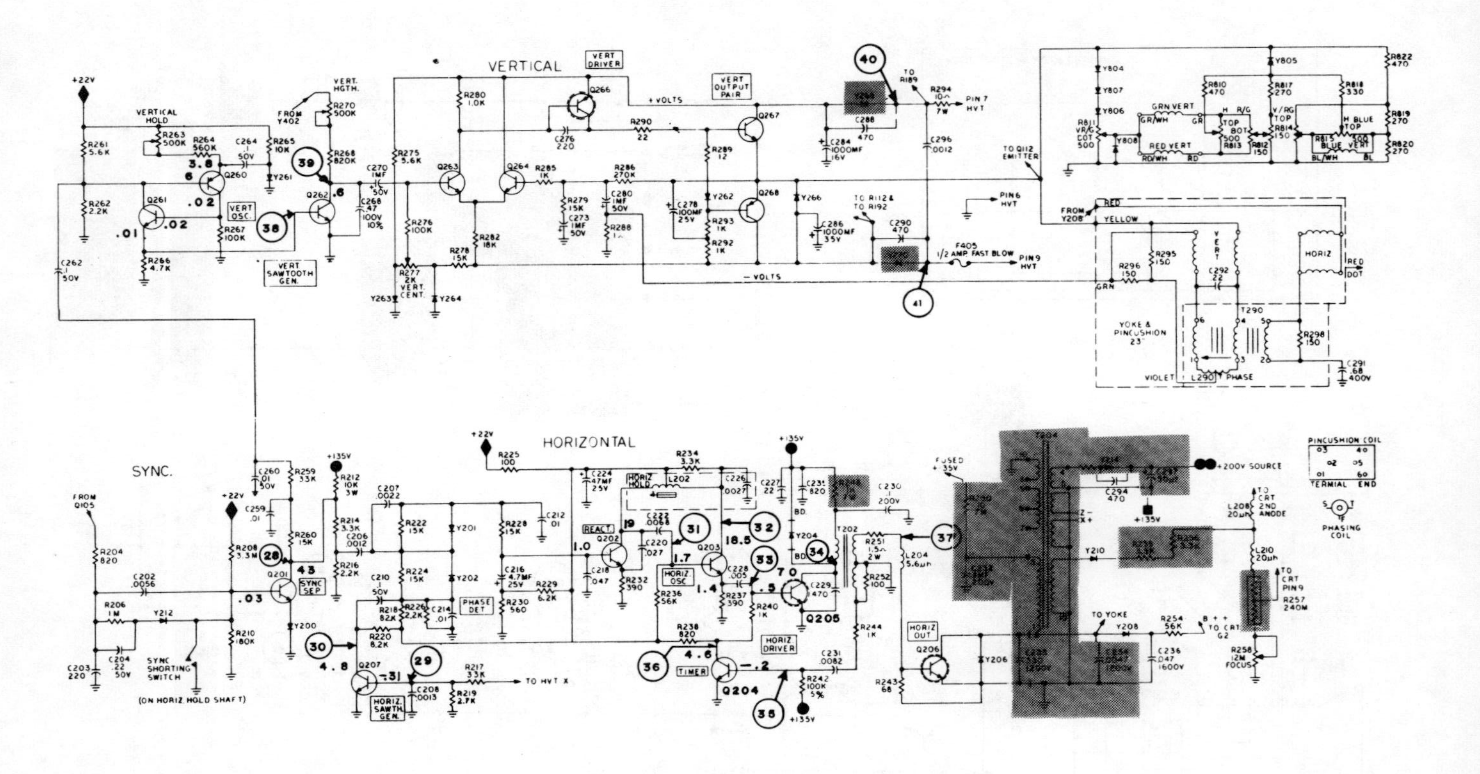
VERTICAL
HORIZONTAL
SYNC.
VERT DRIVER
VERT OUTPUT PAIR
VERT OSC.
VERT SAWTOOTH GEN.
HORIZ. OSC
HORIZ DRIVER
HORIZ OUT
TIMER
PHASE DET
SYNC SEP
HORIZ. SAWTH. GEN.
YOKE & PINCUSHION 23"
SYNC SHORTING SWITCH
(ON HORIZ HOLD SHAFT)
PINCUSHION COIL
PHASING COIL

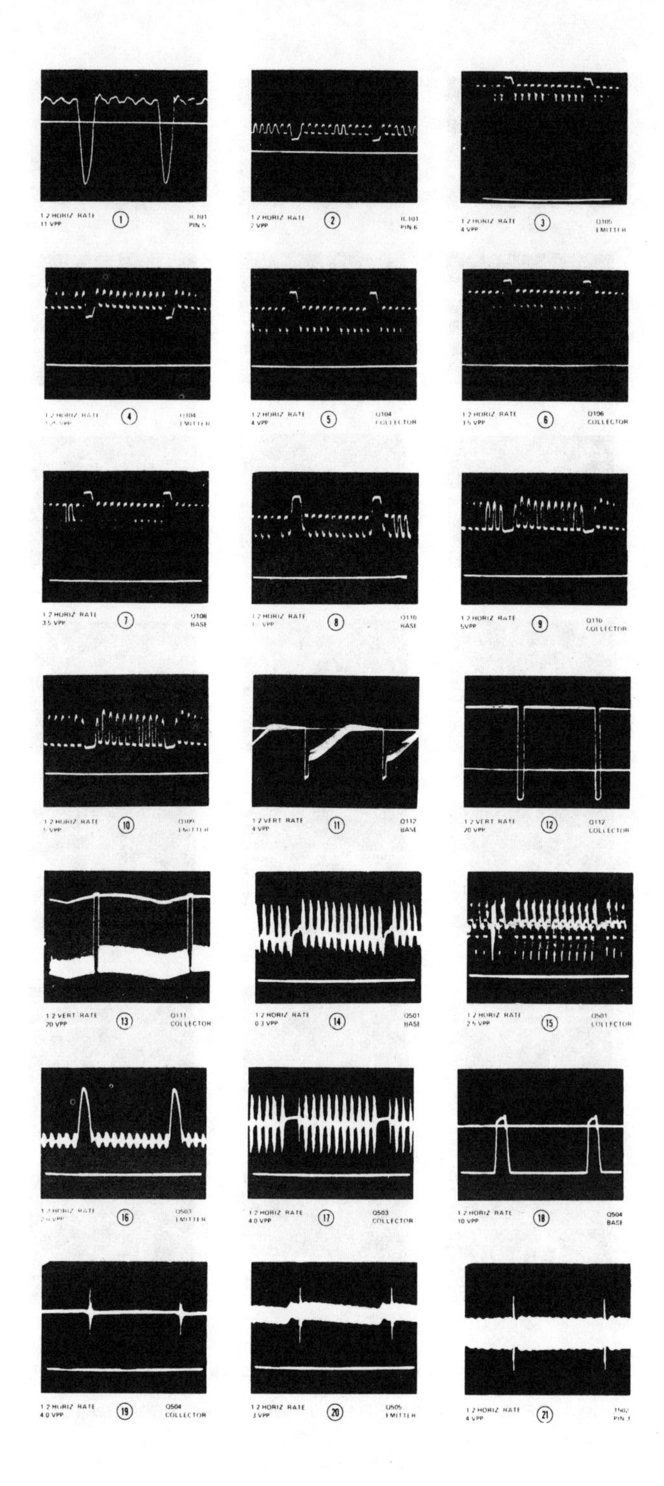

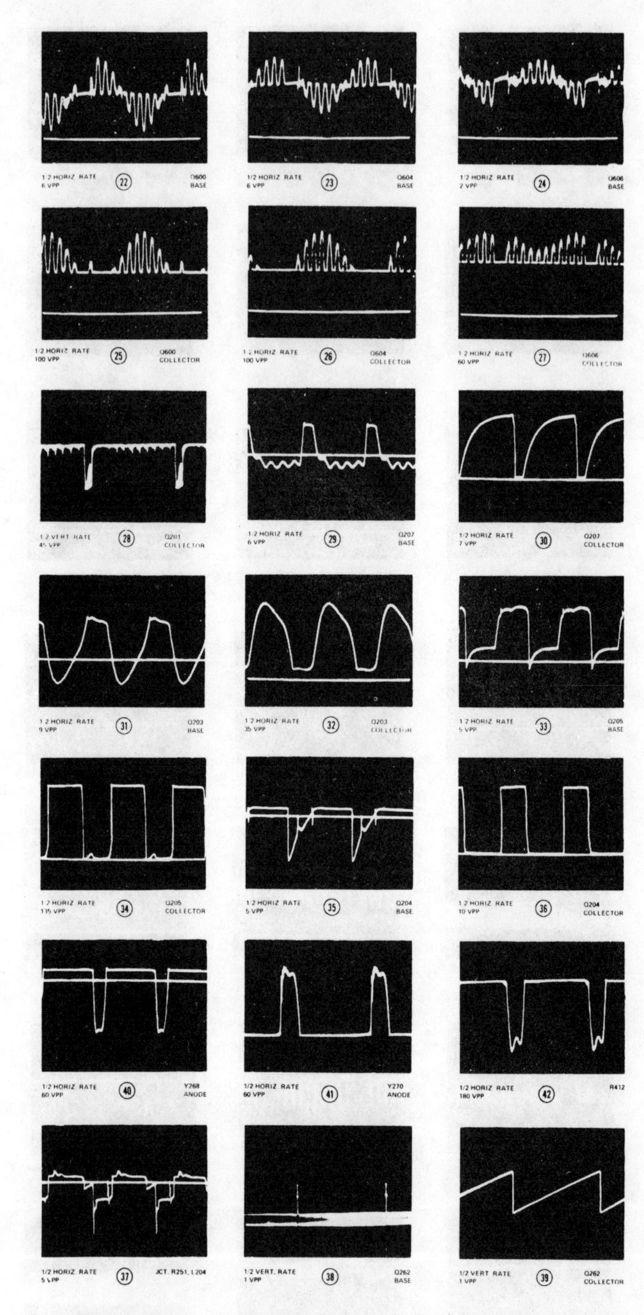

WAVEFORMS FOR POINTS INDICATED
(WAVEFORMS TAKEN WITH KEYED RAINBOW GENERATOR)

OTHER TAB BOOKS BY ART MARGOLIS

Solid-State Circuit Troubleshooting Guide (No. 607)

199 Electronic Test & Alignment Techniques (No. 593)

Modern Radio Repair Techniques (No. 580)

Philco Monochrome TV Service Manual (No. 564)

199 TV Tough-Dog Problems Solved (No. 559)

TV Trouble Diagnosis Made Easy (No. 544)

101 TV Troubles—From Symptom to Repair (No. 507)

TV Servicing Guidebook: Problems & Solutions (No. 484)

INDEX

A

AB classes 92
Absorption circuit, frequency 112
Active filter circuit failure 102
Active filtering circuits 101
Age amplifier circuit 173
Age delay circuit 174
Age keyer circuit 172
Agc-sync configuration 171
AM detection, time constant 109
Amplification 56
Amplifier,
—circuit, agc 173
—circuit, burst 165
—circuit, rf 104
—class A 81
—class A transistor 83
—class B audio push-pull 84
—grounded-base 63
—grounded-grid 62
—limiter 126
—overload complications, rf 106
—trouble, class A 82,84
Amplifiers, audio 88
Amplifiers, other classes 89
Antenna circuit, half-wave 178
Audio amplifier,
—class A 81
—class A transistor 83
—class B push-pull 84
Audio amplifiers, parallel 88
Audio output, transformerless 177
Automatic brightness circuit 181
Automatic degaussing circuit 179
Autotransformer circuit 29
Autotransformer failure 29

B

B-plus, high 70
B-plus, low 72
B-plus, no 69
Bias 54
Bias, junction 43
Blanking, horizontal 160
Blanking, vertical 158
Bridge 129
Bridge circuit, proximity 132
Brightness circuit, automatic 181
Burst amplifier circuit 165
Bypass capacitor circuit 38
Bypass failure 38

C

Capacitance, electronic 96
Capacitor circuit, bypass 38
Capacitor circuit failure 23
Capacitor tester (dip meter) 114
Carbon microphone circuit 117
Cascode (push-pull), class B 87
Cat-whisker diode circuit 47
Cathode-follower circuit 64
Choke circuit 26
Choke failure 28
Class A,
—amplifier trouble 82,84
—transistor audio amplifier 83
—tube audio amplifier 81
Class B,
—cascode (push-pull) 87
—push-pull audio amplifier 84
—push-pull trouble 85
—tapped speaker (push-pull) 87
—transformer-coupled (push-pull) transistor 86
Color,
—demodulator circuits 167
—difference circuits 168
—i-f circuit 162
—killer circuit 163
—oscillator circuit 166
—phase detector circuit 165
—TV reception 161
Code practice oscillator 93
Coil circuit, peaking 37
Coil failure, peaking 37
Configuration, agc-sync 171
Convergence circuit, horizontal 150
Convergence circuit, vertical 151
Conversion circuit 107
Coupling circuits, FET 80
CRT gun failure 147
CRT gun structure 146
Crystal oscillators 110
Current gain 59,62
Cutoff test, transistor 74
CW monitor 116

D

DC restorer 160
Debugging circuit 136
Deemphasis circuit 125
Deflection yoke circuit 148
Degaussing circuit 179
Delay circuit, agc 174
Demodulator circuits, color 167
Detection, time constant 109
Detector circuit, ratio 127
Dimmer circuit 143
Diode circuit,
—cat-whisker 47
—solid-state 41
—zener 44
Diode current, vacuum tube 40
Diode failure,
—germanium 48
—(solid-state) 44
—(vacuum tube) 40
Diode mixer circuits 106
Dip meter circuit 113
Direct conversion circuit 107
Discriminator circuit 123
Distortion 91

Doubler, voltage 99,100
Driver circuit,
noise gate 176
Driver, horizontal 156
Dynamic microphone
circuit 118

E

Efficiency 92
Electrolytic 22
Electron flow 14
Emitter-follower
circuit 66
Envelope, modulation 119

F

Failure, shadow mask 148
FET coupling circuits 80
FET proximity
circuit 133
Filament circuit failure,
—parallel 22
—series 20
Filter,
—circuit 23
—circuit failure 24
—circuit failure,
active 102
—shocker 22
Filtering circuits,
active 101
First i-f circuit 169
Flasher circuit 145
Flow,
—charts 182
Flow charts 182
—electron 14
—signal 16
Flyback circuit 33
Flyback failure 34
FM and PM modu-
lation circuit 121
FM bug 135,138
Frequency 130
—absorption circuit 112
—multiplier 111
—resonant 36

G

Gain, current 59,62
—power 62
—voltage 59,62
Germanium diode
failure 48
Grounded-base amplifier 63
Grounded-cathode
circuit 58
—failure 60
Grounded-emitter
circuit 60
Grounded-grid
amplifier 62
Gun failure, CRT 147
Gun structure,
CRT 146

H

Half-wave antenna
circuit 178
Heater circuit,
series 19
Heating element 18
Horizontal blanking 160
—convergence circuit 150
—driver 156
—oscillator 155
—output 157

I

I-f circuit, color 160
—first 169
—second 170
—third 170
Impedance-coupled
circuit 77

Impedance-coupling circuit failure 79
Impedance-matching circuit 35
Incorrect transistor voltages 76
Inductance (reactance tube), 94
Inductance tester (dip meter) 115
Input signal 58,61

J

Jammer circuit 136
Junction bias 43

K

Keyer circuit, agc 172
Killer circuit, color 163

L

Light bulb circuit 39
—failure 39
Light dimmer circuit 143
Limiter amplifier 126
Load 67
Load adjuster, SCR 142
LR time-constant circuit (flyback) 33

M

Metal detector circuit 128
Meter circuit, dip 113
Microphone circuit, carbon 117
—dynamic 118
Mixer circuits 106
Model speed circuit 146
Modulation circuit, FM and PM 121
—envelope 119
—plate 119
—reactance 122
—screen 120
Monitor, CW 116
Monode 39
Multiplier, frequency 111

N

Neutralizing circuit 97
Noise gate circuit 175
Noise gate driver circuit 176

O

Open transistor 76
Oscillator circuit, color 166
—code practice 93
—horizontal 155
—vertical 154
Oscillators, crystal 110
Output, audio 177
—horizontal 157
—transformer failure 36
—vertical 154
Overload complications, rf amplifier 106

P

Parallel audio amplifiers 88
Parallel-tube heater circuit 21
Peaking coil circuit 37
Peaking coil failure 37
Pentode circuit 50
Phase detector circuit, color 165
Photo detector circuit 140
Plate modulation 119
PM and FM modulation circuit 121

Power gain 62
Preemphasis circuit 125
Properties, circuit 13
Protection circuit, rectifier 103
Proximity bridge circuit 132
Proximity circuit, FET 133
Proximity detector circuit 131
Push-pull audio amplifier, class B 84
Push-pull trouble, class B 85

R

Ratio detector circuit 127
RC circuit failure 26
Reactance circuits 35
Reactance modulation 122
Reception, color TV 161
Rectification 40
Rectifier protection circuit 103
Resistance circuit failure 18
Resistance-coupled circuit tests 68
—transistor circuit 74
—tube circuit 67
Resistances 62
Resonant frequency 36
Restorer, DC 160
Rf amplifier circuit, transmitter 109
—overload complications 106
—receiver circuit 104
Rf circuit tests 105
Rf power amplifier circuit 108
Rf transformer 31
—failure 32

S

SCR load adjuster 142
SCR switch circuit 141
Screen modulation 120
Second i-f circuit 170
Sensitivity 129
Separator circuit, sync 175
Series filament circuit failure 20
Series tube heater circuit 19
Shadow mask failure 148
Shadow mask structure 147
Shocker 22
Signal flow 16
Signal, input 58,61
Siren circuit 134
Speaker (push-pull), class B 87
Speed circuit, model car 146
Stepup, voltage 98
Structure, shadow mask 147
Sweep, vertical 152
Switch circuit, SCR 141
Sync separator circuit 175

T

Tank circuit 36
Tapped speaker (push-pull), class B 87
Telephone FM bug 138
Testing electron flow 14
Tetrode circuit 50
Tetrode failure 52
Thermal runaway 58

Third i-f circuit 170
Time-constant, AM detection 109
Time-constant circuit 25
—LR 33
Transducer (microphone) circuit 118
Transformer circuit 30
—coupled circuit 79
—coupled (push-pull) transistor, class B 86
—failure 30
—failure, output 36
—failure, rf 32
—rf 31
Transformerless audio output 177
Transistor audio amplifier, class A 83
—circuit 52
—circuit, resistance-coupled 74
—class B transformer-coupled (push-pull) 86
—cutoff test 74
Transmitter rf amplifier tube circuit 108
Trigger circuit 130
Triode 48
Triode failure 50,52
Tube heater circuit 19
Tuning indicator 139

V

Vacuum tube circuit, pentode 50
—circuit, tetrode 50
—diode circuit 40
—triode 48
Varactor circuit 46
Varactor failure 47
Vertical blanking 158
—convergence circuit 151
—oscillator 154
—output 154
—sweep 152
Voltage doubler 99,100
Voltage gain 59,62
Voltage stepup 98
Voltages, incorrect 76

Y

Yoke circuit, deflection 148
Yoke failure 149

Z

Zener diode circuit 44
Zener failure 46

This book was created by
Tony Potter Publishing Ltd.

Edited by: Sheila Mortimer and Deirdre Rennison Kunz
Text layouts: Sally Symes and Fran Rawlinson
Art Director: Tony Potter

Published in Ireland by RíRá, an imprint of
Gill & Macmillan Ltd
Hume Avenue, Park West, Dublin 12
with associated companies throughout the world
www.gillmacmillan.ie

ISBN 0 7171 3532 2

Printed in China

Irish Jokes

Compiled by Duncan Crosbie

Illustrated by Peter Rutherford

Fergal got a job driving a one-man bus. One day there was a terrible crash and the garda came to investigate. When asked what had happened, Fergal replied, "How should I know, I was away upstairs collecting fares at the time."

Two drunks were walking home along the railway line. "This is a heck of a long flight of stairs," slurred one. "It's not the number of steps that worries me," said the second, "it's the low railings."

"Excuse me, sir. The Invisible Man's outside."
"Tell him I can't see him."

"We're a man short tomorrow," said the foreman.
"Well, why don't you call my brother," said Walsh. "He can do the work of two men."
"OK," said the foreman, "tell him he starts tomorrow – and you're fired."

Eamonn and Anthony were flying home to Dublin from London, when the captain came on the intercom.
"Ladies and gentlemen, one of the two engines has failed and we'll be two hours late arriving in Dublin."
Eamonn turned to Anthony and said,
"I sure hope the other engine doesn't go or we'll be up here all night!"

Seamus went to visit his doctor.
"Doctor, doctor, I've got two things I must ask you. Could I possibly have fallen in love with an elephant?"
"Of course not," said the doctor. "What's the other question?"
"Do you know anyone who wants to buy a very large engagement ring?"

A dangerous criminal had escaped, so the Dublin garda sent the usual photographs all round Ireland – three pictures of the wanted man, from the left, from the front and from the right. After a few days, they got a message from a detective in Cork:
"Have captured the fellow on the left, and the one in the middle. At the rate I'm going I'll soon get the fellow on the right as well."

Dominic was seeing the doctor about his broken finger.
"When it heals, will I be able to play the piano, doctor?"
"Yes, of course, you will, Dominic."
"Oh, that's great, doctor. I never could before."

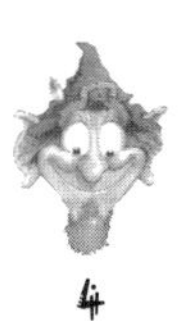

☘ A tourist was driving down a lane in Kerry when a hay cart suddenly emerged from a field right in front of him. Unable to stop, he swerved off the road into the field, where the car turned over.
"Did you see that?" said one Kerryman to the other. "Some of these tourists are terrible drivers. We only just got out of that field in time."

☘ Doherty and Doyler were flying over the Sahara Desert. Doherty looked out of the window and said to Doyler: "Will you look at all that sand there! I wonder what they're going to build when the cement arrives."

☘ A smartly dressed woman got on a crowded bus in the middle of Galway. The only seat was next to a scruffy wee boy, who kept on sniffling. At last, the woman could stand it no longer.
"Have you not got a hanky?" she asked.
"Yes, but me mammy told me not to lend it to strangers."

Donal was up in the attic when he found a dirty old lamp. He thought he'd give it a clean, but when he began to polish it, a genie appeared in a great flash.
"Master," said the genie, "I will grant you three wishes. Anything you want, I will provide."
Donal thought for a moment, and said, "Bring me a big bottle of fizzy lemonade!"
At once, the genie gave him a large bottle of lemonade, and said, "Here you are, master. Drink as much as you like, and it will never become empty."
Donal began to drink and, sure enough, as soon as he had drunk the lemonade, the bottle was full again.
"Hey, that's great!" said Donal.
"And now, master, what are your second and third wishes?" asked the genie.
"You can bring me two more bottles of that lemonade," said Donal.

Aidan was digging a hole when his friend walked by.
"What are you doing, Aidan?" he asked.
"I'm digging a hole to bury my old dog, who's died," replied Aidan.
"So why are you digging these other holes?" said his friend, pointing to three other holes nearby.
"They were for the dog," answered Aidan, "but they weren't big enough."

"I'm the unluckiest person in the world," Aoife sighed. "I bought a non-stick pan yesterday and I can't get the label off."

The Irish invented bottles with small necks so they wouldn't fall in and drown.

How do you make a cigarette lighter?
Take out some of the tobacco.

Finnegan tripped over the edge of the mat and, as he fell, he caught the point of his chin on the piano keyboard and knocked himself out. When he came round, his wife asked, "Whatever happened, Finnegan? Did someone hit you? Who was it?"
"I don't know," said Finnegan, "but whoever did it, he had a fine set of teeth."

A garda stopped Paddy and Seamus in the street.
"What's your name and where are you from?" he asked.
"I'm Paddy O'Reilly of no fixed address," said the first.
"And I'm Seamus O'Toole and I live in the flat above Paddy."

Where does Dracula stay when he goes to New York?
The Vampire State Building.

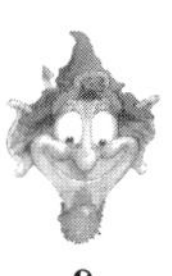

☘ Watch out for the sign on the coast road by a small Sligo village:
"When you can't read this sign, the road is flooded."

☘ "I don't mind dying," said Cian. "It's just that you feel so stiff the day after!"

☘ Flynn went to the doctor, who examined him and said, "I want you to come back and see me tomorrow for a cortisone injection."
Flynn rushed home to his wife and said, "You'll never guess. The doctor's going to give us a car tomorrow!"

☘ Rafferty was taken into hospital. The nurse tucked him into a nice clean bed, and said, "Now you're to stay there and not get up at all. Just let me know if you want a bedpan."
"A bedpan!" cried Rafferty. "Are you telling me I have to do my own cooking?"

How do you make a Kerryman burn his ear?
Ring him when he's ironing.

How do you confuse a Kerryman?
Put two shovels against a wall and ask him to take his pick.

How do you keep a Kerryman happy for a whole afternoon?
Write PTO on both sides of a sheet of paper.

How do you make a Kerryman laugh on a Monday?
Tell him a joke on a Friday.

Have you heard about the Kerryman who learned how to cut his fingernails with his left hand?
He was worried he might lose his right hand.

How do you recognise a Kerryman on an oil rig?
He's the one throwing crusts of bread to the helicopters.

How many Kerrymen does it take to paint an upstairs window?
Two. One to paint the window, and one to hold the ladder.
How many Kerrymen does it take to paint a downstairs window?
Twelve. One to paint the window, one to hold the ladder, and ten to dig a hole for the ladder.

Why do Kerry dogs all have flat faces?
From chasing parked cars.

What do you call a Kerryman on a bike?
A dope peddler.

How do you brainwash a Kerryman?
Fill his wellington boots with water.

How do you recognise a Kerry pirate?
He's the one with a patch over each eye.

Have you heard about the Kerryman who drove his new car over the cliff?
He wanted to test the air brakes.

Kerry sergeant to his men in the middle of a battle: "Keep firing, men. Don't let the enemy know we're out of ammunition."

A Donegalman rushed into a barber's shop with a pig under his arm.
"Where did you get that?" asked the barber.
"I won him in a raffle," said the pig.

"How do you like the new doctor, Ciara?"
"He's that nice, he makes you feel really ill."

O'Brien had a job at the circus, but when he got home one night he remembered he'd left the door of the lion's cage open.
"What does it matter?" he said to himself. "Nobody's going to steal a lion."

Aunt: "How did John do in his history exam?"
Mother: "Badly. But it wasn't his fault. All the questions were about things which happened before he was born."

City teacher: "If there were six sheep in a field, and one jumped over the gate, how many would be left in the field?"
Country pupil: "None, sir."
City teacher: "The correct answer is five. You obviously know nothing about maths."
Country pupil: "You obviously know nothing about sheep, sir."

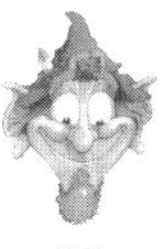

Teacher: "Niall, explain the following words by using them in a short sentence."
Niall:
"*Fascinate* - Sean has nine buttons but can only fascinate.
Rapture - I rapture parcel for you.
Office - My dad fell office chair.
Venom - I don't know venom going to the circus.
Juicy - Juicy that boy over there?"

Teacher: "The River Lee flows into Cork harbour so we call that its mouth. Can you tell me where the source is?"
Niall: "At the other end, sir."

Teacher: "Tell me your three favourite things about school, Niall."
Niall: "Christmas holidays, Easter holidays and summer holidays, sir."

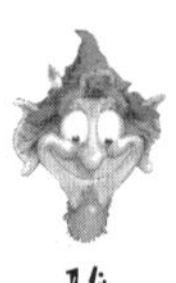

Headmaster: "You're twenty minutes late again, Maire. Don't you know what time we start work at this school?"
Maire: "No, sir, you're always hard at it when I arrive."

Teacher: "Niall, why have you just swallowed that 50 cent piece?"
Niall: "It's my lunch money, sir."

Teacher: "Name four things that contain milk."
Niall: "Butter, cheese and, and ... two cows."

Teacher: "Niall, cleanliness is next to what?"
Niall: "Impossible, sir."

Teacher: "Cillian, where have you been? You should have been here at nine o'clock!"
Cillian: "Why, sir, what happened?"

Teacher: "Why was President de Valera buried in Glasnevin?"
Niall: "Because he was dead, sir."

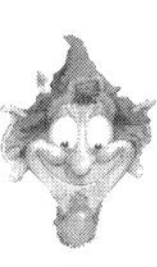

A school inspector was visiting Niall's school. In class, he asked him, "Who was it knocked down the walls of Jericho?"
"It wasn't me, sir," said Niall nervously.
Furious about the low standard in the school, the inspector reported this to the headmaster.
"I asked a young lad called Niall who knocked down the walls of Jericho, and he told me it wasn't him!"
"The young rascal," said the headmaster, "I bet it was him all the time."

What day of the week do Kerry boys and girls play truant from school?
Saturday.

An Irishman living in England went for a job on a building site.
"Can you brew tea?" asked the foreman.
"To be sure I can," answered the Irishman.
"Good. Can you drive a fork lift?" the foreman asked.
The Irishman looked at him and replied: "Why, how big's the teapot?"

Why did the cookie go to the hospital?
It felt crummy.

Why do cows wear bells?
Because their horns don't work.

There was an accident in the Irish Sea today. A ship carrying red paint collided with a ship carrying blue paint. Both crews are now marooned!

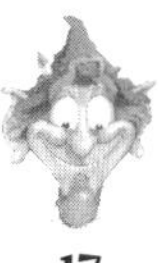

Kevin met Conor walking along the lane with his dog.
"Off for a walk with your dog, are you, Conor?"
"Yes, my dog and I go for a tramp in the wood every day."
"Does the dog enjoy it?"
"Yes, but tramp's getting a bit fed up."

Seamus: "My dog's a mathematical genius."
Liam: "How can you tell?"
Seamus: "Every time I ask him what five minus three minus two is, he says nothing!"

Eamonn: "Our dog doesn't eat meat."
Liam: "Your dog doesn't eat meat? Why not?"
Eamonn: "We don't give him any."

Niall: "My dog's got no nose."
Rick: "How does he smell?"
Niall: "Awful!"

Mrs O'Flaherty goes into the pet shop.
"Have you any dogs going cheap?"
"Sorry, Mrs O'Flaherty, all our dogs go 'Woof!'"

Declan: "My dog chases everyone he sees on a bike. What should I do?"
Aoife: "Take the bike away from him at once."

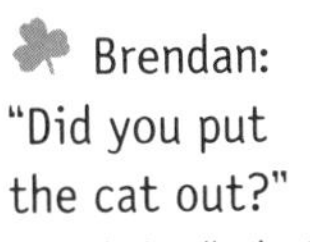

Brendan: "Did you put the cat out?"
Dominic: "Why? Is it on fire?"

Paddy: "My dog can say his name."
Seamus: "Amazing! What is his name?"
Paddy: "Woof!"

Fergal: "There's a black cat in the kitchen."
Maeve: "Don't worry, they're lucky."
Fergal: "This one is. It's eaten your dinner."

"My dog's a blacksmith."
"Prove it."
"Just give him a kick and he'll make a bolt for the door."

Mick: "What's the biggest mouse in the world?"
Fergal: "A hippopotamouse."

Fiachra: "Our dog bit my leg last night."
Dara: "Did you put anything on it?"
Fiachra: "No, he liked it just as it was."

Liam's house caught fire one night, so he phoned the fire brigade and told them to come at once.
"Have you done anything to try and hold the fire back?" asked the fireman.
"Yes," said Liam, "I've been pouring water on it."
"Oh well," said the fireman, "there's no point in coming. That's all we can do."

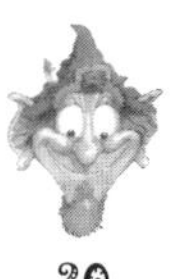

☘ How do you recognise a bath made in Kerry?
It's got taps at each end to keep the water level.

☘ How can you tell a submarine designed by a Kerryman?
It has windows that open.

☘ What do you call a Kerryman who keeps bouncing his head off the wall?
Rick O'Shea.

☘ What do you call a Kerryman who rides his bike on the pavement?
A psychopath.

☘ What do you find at the top of a Kerry ladder?
A STOP sign.

☘ What do you call a Kerryman hanging from the ceiling?
Sean D'Olier.

What do you do if a Kerryman throws a pin at you?
Run like mad - he's probably got the grenade between his teeth.

What do you call a Kerryman under a wheelbarrow?
A mechanic.

Mick saw a notice outside a garda station: "Man Wanted for Robbery". So he went in and applied for the job.

"How's your son Tomás doing?"
"He's at University taking medicine."
"Oh, I am sorry. Is it doing him any good?"

Brendan: "I'm off to see the doctor. I feel a bit giddy."
Tadhg: "Have you got vertigo?"
Brendan: "No, it's just round the corner."

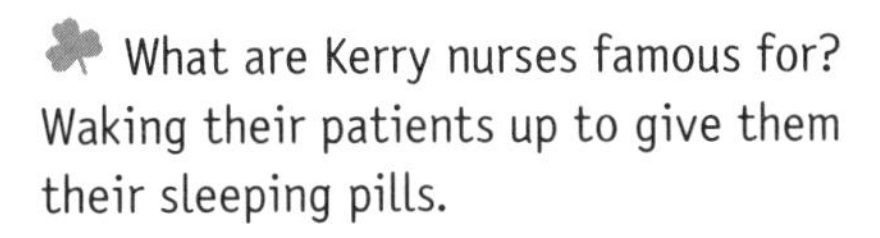

What are Kerry nurses famous for? Waking their patients up to give them their sleeping pills.

An Irish doctor invented a cure for which there was no known disease.

"That does it!" said Shane, coming out of the dentist's after having all his teeth out. "I'll never do that again!"

Have you heard about the Kerryman who had a rope with only one end? He'd cut the other end off.

Then there was the Kerryman who went into an expensive restaurant, ordered a four-course meal, paid for it, and sneaked out without eating it.

Did you hear about the Kerryman who thought Sherlock Holmes was a block of flats?

A Kerry electrician was told to go to a house to mend the broken doorbell. After an hour he came back.
"There's no one in," he said. "I rang the bell four times, but nobody answered."

A Kerryman saw a flat tyre on his car – but he wasn't too worried, because it was only flat at the bottom.

A Kerryman was attacked in the street by a mugger and, after a tremendous battle, finally lost the 50 cent in his purse.
"That was a great fight for only 50 cent," said the mugger.
"Well, I thought you were after the 20 euro I've got hidden in my shoe."

Irish Jokes

How do you recognise a Kerryman in a car-wash?
He's the one sitting on a motorbike.

Did you hear about the Kerryman who got a pair of water skis for his birthday?
He's still looking for a lake with a slope.

During a recent power failure, a Kerryman was stuck on an escalator in a big store for more than an hour.

A Kerryman saw a big sign saying "Keep Death off the Roads". So he drove his car along the footpath.

The Irish invented bagpipes and gave them to the Scots as a joke. The trouble is, the Scots haven't seen the joke yet.

Did you hear about the Irishman who sued the baker for forging the Irishman's signature on a hot cross bun?

Paddy was visiting London, and everywhere he went there were huge crowds. "Is something special happening?" he asked a passer-by.
"Yes, it's the Royal Jubilee," replied the passer-by. Just then there was a loud fanfare of trumpets from Buckingham Palace.
"Jeepers, what's all that?" asked Paddy.
"That's Her Majesty the Queen."
"Is that a fact?" said Paddy. "Well, she sure can play the trumpet!"

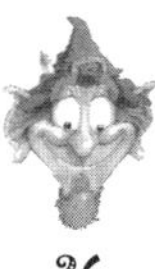

The captain of the ship told Padraig, the mate, to go to cabin 36 because the passenger there had died in the night and was to be buried at sea. An hour later, Padraig reported to the captain: "The man in cabin 26 has been buried at sea as you ordered, sir."
"Idiot! I said 36. Who was in number 26?"
"A man from Cork, sir," Padraig replied.
"Was he dead too?" the captain asked faintly.
"He said he wasn't, sir. But you know what terrible liars these Corkmen are."

Cathal and Sean were on holiday with their parents in a French farmhouse, but they didn't understand a word. In the morning, they were woken by a cock crowing. "Did you hear that?" said Sean. "That's the first word of English we've heard since we got here."

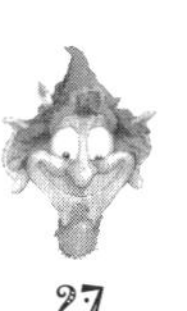

Have you heard about the Kerryman who won a trip to China in a raffle?
He's still there trying to win a trip back.

Paddy was talking to a Spaniard, and asked him what the word *mañana* meant.
"It means," said the Spaniard, "maybe the job will be done tomorrow, or the next day, or the day after. Who cares? Do you have a word for that in Ireland?"
"We do not," replied Paddy. "We don't have a word for anything as urgent as that."

Have you heard about the Kerryman who went to the railway station and asked for a return ticket? When they asked him where to, he said: "Back here of course."

An American tourist was boasting to Jack McCarthy about how the Americans had put a man on the moon.
"That's nothing," said Jack. "The Irish are planning to land a man on the sun."
"Don't be ridiculous," said the tourist. "He'd be burnt to a cinder."
"Oh, we've thought of that," said Jack. "We're sending him at night."

Begorra! I forgot my Sun hat!

Paddy picked up his car keys and went outside to his car, but he couldn't see it anywhere. He rushed back inside and phoned the gardaí.
"Help! Help! My car's been stolen!" said Paddy.
"I see, sir," said the garda. "Where was your car the last time you saw it?"
"Wasn't it on the end of my car keys?" wailed Paddy.

"How far is it to the next village?" asked the tourist. "It's seven miles," the farmer said, "but only five if you run."

Paddy opened the newspaper one day and saw an announcement that he'd died. He quickly phoned his friend O'Keefe.
"Did you see the paper today?" asked Paddy. "It says I died!"
"Yes, I saw it," replied O'Keefe. "Where are you calling from?"

Emmet was boasting about the new clock he'd bought. "It goes eight days without winding."
"And how long does it go if you wind it?" asked Orla.

☘ Austin went into a hardware shop. "I'd like some nails, please," he said.
"Certainly," said the man in the shop, "how long would you like them?"
"Well, if it's all right with you," said Mick, "I'd like them for ever."

☘ Brian phoned Dublin Airport and asked for the information desk. "How long does it take to fly to New York from Dublin?" he enquired.
"Just a moment, sir," replied the clerk.
"Thank you very much," said Brian, and hung up.

☘ An Irish scientist invented the world's strongest glue, but he couldn't get the top off the bottle.

"What are you giving your wife for Christmas?" asked Casey.
"She wants something with diamonds, so I've bought her a pack of cards," said O'Flaherty.

Eugene was dozing on a bench in St Stephen's Green when the park-keeper came up and gave him a dig in the ribs.
"Hey, you! I'm going to close the gates."
"Sure, that's all right," said Eugene. "Mind you don't slam them, won't you."

Two Irishmen met in the street. "Have you seen Mulligan lately?" asked one.
"Well, I have and I haven't," replied the other.
"And what do you mean by that?" the first one said.
"Well, it's like this you see. I saw a chap I thought was Mulligan, and he saw a chap he thought was me. But when we got up to each other - it was neither of us!"

Two tigers were walking down O'Connell Street. "I thought you told me," said one, "that this was a busy place."

Mrs O'Rourke to street musician: "That sounds grand. Do you always play by ear?"
"No, mam. Sometimes I play round the corner."

Mick met Liam walking along carrying a huge lobster. "Are you taking that lobster home for tea?" he asked Liam. "No, he's had his tea," said Liam. "I'm taking him to the pictures."

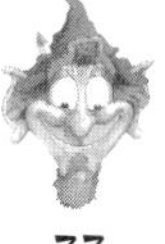

"I ran after you," said Eoin, "but when I caught you up you'd gone."

Kennedy came round the corner and saw Murphy staggering away from his wrecked car.
"Have an accident, Murphy?" he said.
"No thanks, O'Toole – just had one," replied Murphy.

Two tramps met on the banks of the River Liffey in Dublin.
"I saw you outside the Morrison Hotel last night," said one.
"Sure that's where I'm staying," replied the other.
"What, in the Morrison?"
"No, outside."

A garda in Grafton Street went up to a street musician who was playing the guitar.
"Have you got permission to play your guitar in the street?" he asked.
"Well, no, garda, I haven't," said the musician.
"In that case I must ask you to accompany me."
"Certainly, garda. What would you like to sing?"

Mother: "One cake is enough, Niall."
Niall: "That's not fair. You said you wanted me to eat properly, but you won't let me practise."

A nervous tourist was peering over the edge at the Blarney Stone.
"Do people fall off here often?"
"No, sir. Only once."

O'Leary: "How did you find the weather when you went to Limerick?"
Rafferty: "Just opened the door and there it was."

Fergus and his wife went big game hunting in the jungle, but at the end of the day Fergus had only caught a pygmy and a snake. "That's all right," said his wife, "we'll have snake and pygmy pie."

Callaghan took his car for a service but he couldn't get it through the church door.

Irishmen are like bagpipes. You don't get a sound out of them until they're full.

"What would you be if you weren't Irish?" Robert asked Dara.
"Ashamed," said Dara.

"I was going to give that lad a nasty look," said O'Leary, "but he already had one."

What's black and shrivelled and hangs from the ceiling?
A Galway electrician.

Why do Tipperarymen always carry a little rubbish in their pockets?
Identification.

A Wicklowman applied to have his name in *Who's Who*. He was turned down, but they offered to put his picture in *What's This*.

What's red and white and floats upside down in the River Lee?
A Corkman caught telling Kerryman jokes.

How can you tell a Dubliner with an inferiority complex?
He thinks other people are as good as he is.

Have you heard about the Kerryman who damaged his health by drinking milk?
The cow fell on top of him.

Have you heard about the Kerryman who put his television set in the microwave?
He wanted to have a TV dinner.

Have you heard about the Kerry girl who had one arm shorter than the other?
She wanted a job as a shorthand typist.

Have you heard about the Kerry mosquito?
It caught malaria.

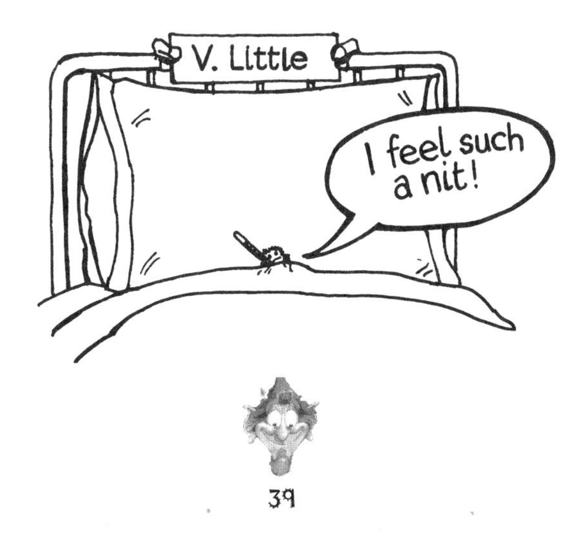

Did you hear about the Kerryman who said he'd give his right arm to be ambidextrous?

Have you heard about the Kerry typist who thought punctuation meant being at the office on time every morning?

Have you heard about the Kerry fire extinguisher factory? It burned to the ground.

Colm: "Where were you going when I saw you coming back?"

"What's wrong with Oisín?" asked Lorcan.
"I don't know, Lorcan. Yesterday he swallowed a spoon and he hasn't stirred since!"

Danny: "That woman looks like Molly Green."
Rory: "You should see her in red."

Did you hear about the Kerryman who was one of the world's great organists?
He had to retire because his monkey died.

Have you heard about the Kerryman who bought a paper shop?
It blew away.

Did you hear about the Kerryman who won the Tour de France?
He set off on a lap of honour and hasn't been seen since.

Have you heard about the Kerryman whose library burned down?
Both his books were destroyed.

Have you heard about the Kerry tap dancer who had to give it up?
He kept falling in the sink.

"Hey, Patrick, that's an odd pair of shoes you're wearing. One black and one brown."
"Not at all. I've got another pair at home just like this."

"Hey, Patrick, why are you wearing your trousers inside out?"
"It's the only pair I've got, and there are holes in the other side."

Ross went round to visit Rory and found him peeling off the wallpaper in the sitting-room.
"I didn't know you were decorating, Rory," he said.
"Decorating?" said Rory. "I'm moving."

A Kerryman went skating on a frozen lake. They warned him the ice was thin.
"Oh, don't you worry about that," he said. "I'll just skate on one foot."

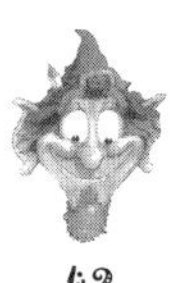

On a hot summer's day, Brady was up a ladder painting his house, wearing two overcoats and perspiring heavily.
"Why are you dressed like that?" asked O'Neill.
"It's the instructions on the can," replied Brady. "It says you're to put on at least two coats."

Fiach: "Why did Padraig Harrington wear two pairs of trousers in the Ryder Cup?"
Brian: "In case he got a hole in one!"

Garda to car driver after a crash: "And whose fault was the accident?"
"I really don't know, garda. I wasn't looking."

Did you hear about the Kerrywoman ironing her curtains?
She fell out the window.

Garda to woman car driver after a crash:
"May I see your driving licence, please?"
"A driving licence! Who'd give *me* one?"

Shane: "Wish I had enough cash to buy Dublin Castle."
Fergal: "Why on earth do you want to buy Dublin Castle?"
Shane: "I don't. I just wish I had the cash."

Mick arrived at work four hours late.
"What's your excuse this time?" asked the boss.
"The trouble with me," said Mick, "is that I sleep slowly."

Danny and his pals were playing together when a big lad came up to them and asked, "Which one of you is Michael O'Shea?"
"I am," said Danny, so the big lad knocked him down. Danny got up grinning, so the big lad knocked him down again, and again Danny got up grinning.
"Why are you laughing?" roared the big lad. "D'you want me to thump you again?"
"The laugh's on you," said Danny. "I'm not Michael O'Shea at all!"

Barra went into Mario's fish and chip shop late one night.
"Have you got any chips left?" he asked.
"Certainly, sir," said Mario.
"Well, it serves you right for frying so many," said Barra.

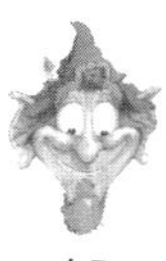

Nurse: "Why do doctors wear masks during operations?"
Student: "It's so that if they make a mistake, no one will know who dunnit."

A fellow walked into a bar in Dublin and asked the barman if he'd heard the latest Kerryman joke.
"I'd better warn you," said the barman, "I come from Kerry myself."
"That's all right," said the fellow, "I'll tell it slowly."

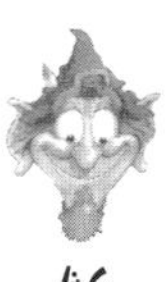

Peadar took a job in a Dublin office as an assistant answering phone calls. One day the phone rang. Peadar picked it up and put it down almost immediately.
"Who was that?" asked the boss.
"Some fool saying it was a long distance from New York. I told him everybody knows that."

A Kerryman was at a concert where a ventriloquist was telling Kerryman jokes.
After a bit, he stood up and shouted, "I'm a Kerryman, and we're not all as stupid as you make out."
"It's only a joke, sir," replied the ventriloquist. "I'm sure Kerrymen have a good sense of humour."
"It's not you I'm talking to," roared the Kerryman. "It's the little fellow sitting on your knee."

Maeve: "I found 10 cents on the floor this morning."
Sean: "It must be mine. I dropped 10 cents this morning."
Maeve: "But I found two 5-cent pieces."
Sean: "Then it must be mine. I heard it break in two when it hit the ground."

If wars were fought with words, Ireland would rule the world!

What's the difference between a Clareman and a bucket of fertiliser?
The bucket!

"How do you save a Limerickman from drowning?"
"I don't know."
"Thank goodness."

Dermot: "Waiter, there's a snail eating my cabbage."
Waiter: "Don't worry, they have very small appetites."

Dermot: "Waiter, I don't like this cheese."
Waiter: "But it's Gruyère."
Dermot: "Well, bring me some that grew somewhere else."

Dermot: "Waiter, there's a fly in my soup."
Waiter: "I'm so sorry. It was supposed to be in the stew."

Dermot: "Waiter, there's a fly in my soup."
Waiter: "Just wait a moment – I'll fetch a spider."

Dermot: "Waiter, there's a fly in my soup."
Waiter: "That's impossible. We don't serve meat on Fridays."

Dermot: "Waiter, what's this fly doing in my soup?"
Waiter: "Looks like the crawl to me."

Dermot: "Waiter, there's a fly swimming in my soup."
Waiter: "Really? There's usually only enough for them to paddle in."

Dermot:
"Waiter, call the manager. I can't eat this terrible food."
Waiter: "He couldn't eat it either, sir."

Dermot: "Waiter, there's a worm on my dinner."
Waiter: "Sir, that's fat."
Dermot: "It certainly is. It's eaten all the meat."

Dermot: "Waiter, this lobster only has one claw."
Waiter: "It must have lost the other one in a fight."
Dermot: "Well bring me the winner, then."

Dermot went into a restaurant and said to the waiter, "I'll have asparagus."
"We don't serve sparrows," said the waiter, "but how did you know my name is Gus?"

"Is that bull safe?" Eamon called to the farmer as he crossed the field where it was grazing.
"Oh yes," called back the farmer, "a lot safer than you are."

Conor met Kevin wheeling a barrow-load of manure along the lane.
"What are you going to do with that, Kevin?" he asked.
"Put it on my rhubarb, Conor."
"That's a change. We put custard on ours."

Breda went to the zoo with her friend.
She saw a sign over the kangaroo's cage which said "A Native of Australia".
"And to think my sister went and married one of those things," she said.

Guard to Kerryman on the train going to Dublin: "Pass farther down the train, please."
"That isn't my father. It's my grandpa."

Doctor to Mrs Malone: "Did that medicine I gave to Mr Malone straighten him out all right?"
Mrs Malone: "Oh, it did fine, doctor. They buried him yesterday."

Tourist: "How far is it to Ballymoe, Paddy?"
Paddy: "How'd you know my name was Paddy?"
Tourist: "I guessed it, my good man, just guessed it."
Paddy: "Then you can guess how far it is to Ballymoe if you're so smart."

☘ Two Kerry carpenters were making a cupboard.
"Hey," said one, "these nails are defective. The heads are on the wrong end."
"Don't be an idiot," said the other one. "They're for the other side of the cupboard."

☘ Have you heard about the Kerryman who wanted to be buried at sea?
Three of his friends were drowned digging the grave.

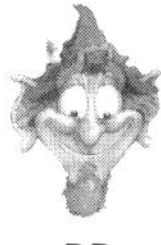

A Kerryman went into a shop and asked for a hundred mothballs.
"But you bought a hundred only yesterday," said the shop assistant.
"I know," said the Kerryman, "but those moths are difficult to hit."

How many Kerrymen does it take to launch a ship? A thousand and one. One to hold the bottle of champagne, and a thousand to bang the ship against it.